Climate Change
and the Agenda for Research

Climate Change
and the Agenda for Research

EDITED BY

Ted Hanisch

Westview Press

BOULDER • SAN FRANCISCO • OXFORD

Copyright © 1994 by Westview Press, Inc.

Published in 1994 in the United States of America by Westview Press, Inc., 5500 Central Avenue, Boulder, Colorado 80301-2877, and in the United Kingdom by Westview Press, 36 Lonsdale Road, Summertown, Oxford OX2 7EW

Library of Congress Cataloging-in-Publication Data
Climate change and the agenda for research / edited by Ted Hanisch.
    p. cm.
    Includes bibliographical references.
    ISBN 0-8133-8791-4
    1. Climactic changes—Environmental aspects.   2. Environmental policy—Developing countries.   3. Greenhouse gases—Environmental aspects.   4. United Nations Conference on Environment and Development (1992 : Rio de Janeiro, Brazil)   I. Hanisch, Ted, 1947–
QC981.8.C5C513   1994
363.73'87—dc20
                                                                                94-13943
                                                                                      CIP

Printed and bound in the United States of America

The paper used in this publication meets the requirements of the American National Standard for Permanence of Paper for Printed Library Materials Z39.48-1984.

10   9   8   7   6   5   4   3   2   1

Contents

About the Contributors

The Center for International Climate and Energy Research-Oslo (CICERO) is a policy research foundation of the University of Oslo, Norway's largest university. Founded by the Norwegian government in 1990, CICERO forms part of a world-wide research effort in support of climate and energy related multi-lateral cooperation.

Ted Hanisch held the position of Director of the Center for International Climate and Energy Research-Oslo (CICERO) until 1993. He is now Deputy Director of the Norwegian Employment Directorate.

Dr. William R. Cline is a Senior Fellow at the Institute for International Economics in Washington, D.C.

Dr. Rudolf Dolzer is a Professor on the Law Faculty of the University of Mannheim, Lehrstuhl für Völkerrecht, in Mannheim, Germany.

Dr. Eunice Ribeiro Durham is a Professor at NUPES – University of Sao Paulo in Brazil.

Dr. Michael Hoel is a Professor of Economics at the University of Oslo.

Dr. Ivar S. A. Isaksen is a Professor of Geophysics at the University of Oslo and Senior Scientist at CICERO.

Dr. Calestous Juma is the Executive Director of the African Center for Technology Studies (ACTS) in Nairobi, Kenya.

Dr. Jane A. Leggett is chief of the Stabilization Branch in the Climate Change Division of the Environmental Protection Agency in Washington, D.C.

Edith Mneney is a Research Associate at Kenya's African Center for Technology Studies (ACTS) Ecotechnology Institute.

Dr. Rajendra K. Pachauri is Director, Tata Energy Research Institute, New Delhi.

Dr. Arild Underdal is currently Prorector of the University of Oslo. He is also a Professor of Political Science at the University of Oslo and Senior Scientist at CICERO.

Dr. T. M. L. Wigley is Director of the Office for Interdisciplinary Earth Studies of the University Corporation for Atmospheric Research, Boulder, Colorado.

1

Introduction

Ted Hanisch

From the early days of the Brundtland Commission until the Rio Summit, policy makers and scientists have struggled with alternative global regimes to prevent dangerous changes in the atmosphere and elements of it. The discussion continued while the UN Intergovernmental Panel on Climate Change (IPCC), from 1989 to 1990, managed to sort out much of the discord in the sciences about climate change itself.

When we at CICERO considered the state of the situation here in Oslo after Rio in late June 1992, we insisted on starting our further work on climate change on the new platform created by the summit. We decided to invite a group of eminent, leading experts to assess the platform created by the UN Framework Convention on Climate Change (FCCC). We wanted to hurry forward on this basis, partly because we had tried to assist negotiators in their efforts with some success, and partly because we simply felt that too much literature had been speculating rather freely about possible climate regimes.

The global effort to prevent dangerous climate change is now well into its first phase of institution building. The platform was created by the UN Convention on Climate Change, signed in Rio de Janeiro in June 1992. Any UN convention of this sort is a framework for building new global institutions.

When negotiations concluded in New York in May 1992, the Climate Convention was criticized as non-binding and empty. As months passed, more experts tended to acknowledge that the Convention initiated a large political process which would last for a couple of decades. Patience is a necessity for all those who spend their time and intellect on global environmental problems.

The convention also unveiled real challenges to the academic community throughout the world. Many leading scientists have for years argued for action to limit emissions and preserve sinks and reservoirs of greenhouse gases. When their best assessment was summarized by the UN Intergovernmental Panel on Climate Change (IPCC) in 1990 the political establishment generally accepted the challenge. But, as before, this raised more questions for the academic community, some of which were answered by 1993, and some which were not. This is true for the science of climate change as such, as well as for the scientific analysis of relevance to regulation and response strategies, i.e. the legal, institutional and economic aspects of the problem.

The CICERO seminar in 1992 focused on the agenda for research given the platform established by the Climate Convention. Until then, the academic community to a large extent had discussed its priorities in light of "possible climate regimes." Since we at CICERO had been actively involved in advising negotiators in the Intergovernmental Negotiating Committee, we were early to recognize the trailbreaking function of the convention.

We knew from studies made by our staff on global and regional environmental negotiations that the building of institutions has an organizational as well as a scientific side. When we look at the structure expected when the convention will enter into force probably by mid-1995, we see that much is needed before we have an institution capable of managing the regulatory process. Greenhouse gas emissions have an enormous number of sources, many of which can only be changed by measures of major economic significance. Conflicts of interest are likely to occur within nations and among nations, notably as targets grow firmer and higher.

From a scientific point of view there is a strong need to improve our understanding and empirical knowledge of the causes of climate change. There seems to be agreement among scientists that our best knowledge is sufficient for governments to decide on the first generation of measures. So-called "no regrets" options should be seen as a low insurance premium, given the dramatic and irreversible changes that may take place.

What is less known is whether we have sufficient knowledge and understanding of how governments should act most effectively. There are ambiguities about how governments should put priority on alternative response options to climate change and about how costs and benefits should be compared when it comes to responding now or later.

In this book a number of leading experts from all over the world offer their assessments on where we stand in the process of managing climate change, given the Climate Convention and our present status of scientific knowledge.

Calestous Juma and Edith Mneney start out with the fundamental issue of transfer of new technologies and the building of capacity in developing countries, particularly in Africa. For any long term and lasting approach to meet the challenge of Climate Change, this very practical issue should be on the agenda.

Dr. Juma and Ms. Mneney point to the major potential of existing technologies in reducing greenhouse gas (GHG) emissions. Their argument is that the problem is rooted in the lack of availability. The main message is that technologies are rarely "transferred from the North to the South." As a rule, communities with expertise in the South, given a technical capital base, are likely to fetch or purchase technology from the North. The issue of patents and licences, so tricky to deal with during multilateral negotiations, seems to be less important for building a basic technical competence. Rather, there is a need for basic technical capital and an institutional setting for it.

The authors also point to the biased focus in many studies of the lack of efficiency of development aid. They point on the other hand to the importance of national responsibilities in developing countries, given the fact that barely 5 percent of their GDP are extracted from foreign transfers.

Jane A. Leggett enters into the issue of monitoring and verification. The FCCC entails provision for National GHG inventories, reporting (providing information) about national policies, scientific cooperation and exchange of scientific data. These components could complement each other and support the efficiency of implementing the FCCC. Collecting this sort of information will be important in deciding whether response options and measures are sufficient and to what extent future revisions of the FCCC will be necessary.

National GHG inventories provide a basis for assessing total global emissions, alternative measures and the effects of them over time. In this area the IPCC and Organization for Economic Cooperation and Development (OECD) have done much of the basic work. Methods for assessing emissions of CO_2 from commercial carbon fuels and for most other GHG's are available. However, for noncommercial fuels and non-energy related CO_2 emissions, data and methods are still underdeveloped.

With regard to information about national policies and measures, methods are not likely to be identical for all countries. One can only opt for comparability and coordination of assumptions built into the model studies of national policy options. Here we have a long way to go.

Dr. Eunice Ribeiro Durham writes in her chapter, "Climate Change Policies in Developing Countries," on the responsibility put on developed countries by the FCCC. The principle of "a common but differentiated responsibility" is a major principle of the convention.

There are major differences in responsibilities and possibilities to participate within the group of developing countries. Dr. Durham argues that the largest four, namely Brazil, Indonesia, China and India, carry the responsibility to select development strategies with a view to the limitation of global environmental threats. In this respect the use of forests as carbon sinks is of intrinsic importance.

Through their decisive role in a global effort to limit emissions of GHG, particularly CO_2, these countries have a very strong bargaining position, as could be observed during the negotiations prior to UNCED. In the years to come this position could be used by governments to attract financial and technical support to choose less of the dangerous anthropogenic emissions than industrial countries in their development.

On the negative side, Dr. Durham notes that foreign interest, particularly multinationals, influence the use of resources in a way that may seem contrary to expressions of national sovereignty. They also often seem to face major problems of enforcing national policies in an effective way.

Dr. Rajendra K. Pachauri makes his approach to the role of developing countries from the energy angle. He sets out to assess how to meet energy needs in major developing countries with lower emissions under the regime of the FCCC.

Dr. Pachauri presents studies in support of a strategy that emanates from two realities: 1) There is an important but limited scope for improvement of energy efficiency in developing countries, and 2) efficiency gains should first take place in the North, where most of the best technology is produced.

Assessing existing studies of the potential of alternative energy sources and emissions of GHGs well into the next century, for Asia in general and India in particular, Dr. Pachauri is able to show that there are no easy shortcuts available. The sort of practical grip that energy experts have is of vital necessity to any realistic effort to limit emissions of GHGs.

On the other hand, Dr. Pachauri argues strongly that there are major opportunities to be identified. He also points to what may be gained by a joint implementation of effective measures by developed and developing countries as provided for in the FCCC. His point is that the elements of fear and skepticism which governments in major developing countries in Asia still face can only be overcome by positive experience of North-South co-operation. It is only natural that developing countries regard removal of poverty as the most important element in any strategy for protecting the global environment.

In "Socially Efficient Abatement of Carbon Emissions," Dr. William R. Cline argues that a large-scale and long-term program for major reductions is actually profitable on the macro level. The major contribution from Cline is his insistence on including in his calculations the long-term dramatic changes in the atmosphere due to maximum concentrations of GHGs and their social costs. His major criticism of previous studies is that the often-used time horizon of 50 years is systematically misleading and it draws our attention away from the major threats.

For this reason Cline uses a 300-year horizon in his calculations, proving dramatically the increase in damage costs of climate change and thereby the benefits of mitigation. On the other hand, this long horizon underlines how vulnerable the studies are to the right assessment of future technological change. It is hard to know what technologies we will have available by the year 2100 and later.

Recognizing the imminent difficulties of all such very long term studies, Dr. Cline's chapter is very relevant and valuable input to the sort of broad discussions policy makers need to go through while developing the practical steps in their national strategies and their cooperative efforts.

In the chapter by Dr. Arild Underdal, and in the three subsequent chapters we enter into issues more relevant to what has been achieved by the FCCC as such and what the Parties to the convention might achieve throughout this decade and the next.

Dr. Underdal raises the discussion: What is more important, the legal commitments to limit emissions of the UNCED and the FCCC as they now stand or the process of communication and cooperation? His argument is that international attempts to manage problems do have positive effects, regardless of the substantive decisions which they entail.

He continues by arguing that, first, the solving of problems is more than just making decisions. It also implies learning, identification and diagnosing of new phenomena. Second, joint or coordinated efforts are only one part of the total "climate policy." Unilateral decisions will always encourage international process. Third, nations respond to environmental problems not only through their governments; other important channels of influence are producers' and consumers' behavior.

When considering what matters, it is also necessary to ask what commitments are really to be found in the FCCC. Dr. Rudolf Dolzer carefully examines the text and his conclusion may be surprising to those who only read the headlines in May and June 1992.

As Dr. Dolzer points out, the contention that the FCCC does not entail any substantive commitments is simply misleading. The signatories are committed legally to strive for certain goals, i.e. reduce the level of GHG-emissions by the year 2000 to the level of 1990. Further, the parties are pledged to adhere to certain principles in the policies, particularly the precautionary principle.

Commitments in the FCCC are also pointed out in Jane A. Leggett's chapter in the discussion on procedures for providing reports and scientific information. The developed country parties are also dedicated to covering the full and agreed incremental costs of the commitments taken on by developing country parties.

Dr. Dolzer's conclusion is that the commitments, even though legally binding, are open for interpretation in a way that may reduce their short-term effect. On the other hand, the FCCC is bound to have a bearing on future policies and certainly its principles will be guidelines for future amendments of the FCCC.

The questions then arise: What can governments do to set out a strategically sensible set of measures? Do we have methods to assess what are the right priorities?

Michael Hoel and Ivar S. A. Isaksen enter into this very complicated matter by presenting a model of an alternative to the so-called GWP index (index of the global warming potential of each greenhouse gas).

They argue that a cost effective policy should take all gases into account, a comprehensive approach. While the GWP developed and used by the IPCC only calculates the direct effect of each gas, with a view to their life cycle in the atmosphere, the new model calculates the damage effect measured by increased temperature, with the possibility of testing alternative discount rates. The more weight we put on the interest of future generations, the more we should concentrate on gases with a long lifetime in the atmosphere.

The results of the calculations differ to some extent from what emerges from the IPCC GWP-index. When their method is improved, one should think it would be valuable for a practical approach to the difficult decisions on priority. In one respect it may seem easy to agree that "the problem must be solved." With limited resources, as usual it is not easy to decide what steps to take next year and why not some other measures?

The other important issue for international cooperation which is discussed by the editor is the importance of burden sharing within the group of signatories. During the pre-UNCED negotiations most of the discussion centered on the North-South dimension of burden sharing. Since developed countries will have to carry most of the costs in the foreseeable future, the sharing of burden within the OECD group will in reality prove more important.

The argument here is that the FCCC does not fully recognize the way differences in energy systems, resource bases and industrial structure are decisive for the relative cost of numerically similar limits to emissions. Countries with relatively old-fashioned energy systems will have much lower costs for a certain target than those with more modern systems or countries with a high proportion of renewable energy.

If one combines the "polluter pays" principle with the "grandfathering" principle (i.e. same reduction in percent for all OECD countries), the FCCC will punish those parties who in the past have done the most to limit GHG emissions and acid rain, and support those who have done the least. For further progress, improvement of this weakness is important.

To some extent the unresolved problems of burden sharing can be overcome by the mechanism for "joint implementation of commitments," provided for in the FCCC. By this mechanism of governing, countries where marginal costs are among the highest can choose to cooperate with countries where low-cost alternatives are available and implement some of their measures abroad.

Dr. T. M. L. Wigley makes an assessment of the state of the art of our knowledge of the different greenhouse gases and their contribution to climate change.

First, Wigley discusses the carbon cycle and what uncertainties we have to face. He uses a carbon cycle model to assess what importance the imbalance (the missing sink) actually has with regard to calculation of future CO_2 concentrations and changes in temperature. His conclusion is that carbon storage actually may be larger than calculated so far, due to "CO_2 fertilization" of biomass growth. The size of this effect is, however, difficult to ascertain, and, because of that, calculation of future CO_2 concentrations is very uncertain.

In his second contribution Wigley sets out to assess what holes in our knowledge we should try to eliminate first; where it is most important to know more. Wigley makes reference to studies which show that reducing uncertainty may be more profitable as a measure than reduction of emissions beyond the no-regrets limit.

I can only hope that these initial remarks sharpen the reader's appetite for reading the chapters themselves. At CICERO we were all confident that contributing to the science and policy of handling climate change is a task only for the patient among us.

With this book we hope to have been able to combine solid scientific knowledge with a practical understanding of realistic policy options. At the same time we have collected research from experts with a firm belief in a common responsibility.

I want to thank all of the authors for their contribution, the staff at CICERO for their ongoing support of this venture, Svanhild Blakstad for her perseverance in preparing the manuscript, and Peggy Simcic Brønn for her guidance in managing this project. Thanks to the editors at Westview for their support in this truly international effort.

2

Environmentally Sound Technology Transfer and Capacity Building in Africa: Strengthening Incentive Systems

Calestous Juma
Edith Mneney

Introduction

The use of environmentally sound technologies has been recognized as imperative in enhancing sustainable development. Chapter 34 of Agenda 21 which deals with environmentally sound technology stresses the:

> need for favourable access to and transfer of environmentally sound technologies, in particular to developing countries, through supportive measures that promote technology cooperation and that should enable transfer of necessary technological know-how as well as building up of economic, technical, and managerial capabilities for the efficient use and further development of transferred technology.

The aim of this chapter is to assess the incentive regimes for the transfer and development of technological capacity in Africa with specific reference to environmentally sound technologies. It argues that much emphasis has been put on the international obstacles to the transfer of environmentally sound technologies, but the role of appropriate incentive regimes in the developing countries to facilitate technology acquisition and development has been downplayed. The chapter proposes a range of policy incentives that could be applied to promote environmentally sound technology development in Africa. It stresses that the effective application of the proposed incentives can only be achieved if implemented within a policy framework that puts technology at the heart of the sustainable development process.

Technology and Sustainable Development

The New Technological Awakening

The environmental activists of the 1960s perceived large-scale industrial development as the main source of environmental degradation. As a result, technological development was sometimes seen as a key culprit and an "anti-technology" attitude developed among environmental activists. The work of E.F. Schumacher in the 1970s promoting "appropriate technology" provided the activists with alternative perceptions on technological change.[2]

The notion of technological appropriateness is now being used widely when discussing the relationship between technology and environment. There is a new awakening that industrial development is not only vital in improving the economic performance in developing countries, but the transition away from the current dependence on basic agricultural production (with the related destruction of the ecological base), will require a shift towards industrialization and the related service sector activities.[3]

Most developing countries rely on a narrow range of economic activities and there is widespread recognition that they are unlikely to achieve any significant economic growth unless they diversify their economies. It has been argued that the adaptive policies required by the developing countries to deal with the long-term effects of climatic change are the same as those needed to improve economic stability.[4]

Industrial production, therefore, if undertaken in a sustainable manner, could lessen the pressure on bioproductive resources by offering alternative employment.[5] The basic technical possibilities for making this transition already exist. Whether these options are adopted will depend largely on the policies and practices as well as the range incentives available in these countries to promote technological development.

In the 1990s, with particular impetus from Our Common Future as well as the results of the United Nations Conference on Environment and Development (UNCED), the world community has recognized the importance of technological innovation in responding to environmental problems. This has made it possible for the private sector, mainly in the industrialized world, to engage in the promotion of the sustainable development agenda.[6]

Technological innovation, which used to be seen largely as a threat to the environment, now offers new opportunities for reducing environmental degradation and promoting sustainable development.[7] This factor has changed the terms of relations between the industrialized and developing countries. In the 1970s, technology transfer was seen as a potential threat to the environment, thereby requiring regulation and control. In the 1990s, technology transfer is now being seen as a source of opportunities for promoting sustainable development. The challenge therefore is how to move from the traditional control of technological flow to new approaches of technology assessment that take environmental concerns into consideration.[8] The transition towards greater application of environmentally sound technologies is being mediated mainly through research and development (R&D) activities.

Research and Development

Probably the most important environmental development in the industrialized countries in the 1990s is the changing character of environmental research. In the 1970s and 80s, environmental research was largely disciplinary and piece-meal in character. More recently, research activities have become increasingly multi-disciplinary and systemic in approach. The role of social science research is increasing.

Despite these changes, official environmental R&D funding in the OECD countries remained low in the early 1990s, often not more than 3% of total government expenditure.[9] This figure, however, covers only explicit environmental R&D. Because of the interdisciplinary nature of environmental R&D, funding continues to go to projects that indirectly relate to environment. Such implicit environmental R&D, though difficult to quantify, will continue to rise in view of the fact that more research projects are starting to incorporate environmental concerns into their activities.

On the whole, environmental R&D expenditure has remained low in relation to the degree of concern over environmental issues. Support has tended to be national, yet most of the major problems require international research efforts. Part of this may be explained by the fact that most of the research policies formulated in the industrialized countries in the 1980s treated research as a tool for international competition and therefore focused on areas with obvious national benefit.[10]

It is notable that the share of private sector funding for environmental R&D in the OECD countries is estimated at 80%, compared to 50–60% for total OECD R&D.[11] This is explained by the shift towards product-oriented environmental R&D as well as the impact of two decades of environmental regulation. The rise of the "environment industry" devoted to clean-up operations is on the rise, with a global turn-over of US$200 billion. This market is expected to reach US$300 billion by the year 2000. At this rate, the environment industry is comparable to the pharmaceutical and aerospace industries, but with higher prospects for growth.

One of the most significant developments in research has been the shift from "clean-up" approaches towards more comprehensive preventive technological innovations. This reshaping of technology has led to more research investment in low-emission and low-waste technologies – a category that is increasingly being referred to as "cleaner technologies". Political considerations, regulatory measures as well as public awareness are being brought to bear on the direction of technological innovation.

The intensification of basic environmental research is also accompanied by measures to promote the commercialization of such technologies. New institutions are being established, especially in OECD countries, to promote innovation in environmental technologies. In addition, support for collaborative programmes is also increasing, especially in the EEC countries, whose annual expenditure on environmental R&D averaged US$43 million over the 1981–88 period (or 7.6% of its total R&D budget).

Regional programmes such as the EEC's Science and Technology for Environmental Protection (STEP), European Programme on Climatology and Natural Hazards (EPOCH) and the Joint Opportunity for Unconventional or Long-term Energy Supply (JOULE) are taking root. These changes, however, should not conceal the fact that, in many countries, funding for research has not expanded enough to meet economic challenges. In most areas, there has been a shift from basic research towards applied research. It has also been argued that the decline in public sector R&D may undermine support for basic research in areas which do not show obvious and short-term commercial application.

It is expected that technological cooperation will be extended to the eastern European countries through bilateral programmes. However, cooperation with the developing countries is likely to occur largely in the field of the management of natural resources and the transfer of specific environmentally sound technologies. No major technological cooperation programmes are envisaged, except in areas covered by international conventions on issues relating to climatic change. It is notable that it has been very difficult to reach viable arrangements on issues relating to technology transfer during the biodiversity convention negotiations.

In the more advanced developing countries, emphasis seems to be placed at two main levels: the development of environmental research and training institutions and support for incremental technological innovations at the firm level. It is expected that training in key areas of environmental research will continue to be a major theme in the developing countries.[12] This is mainly because the basic scientific knowledge required to embark on significant programmes in technological development or natural resource conservation is still lacking.

But where such capacity exists, the countries face major challenges in providing the institutional basis for the utilization of such capacity. Institutional development as well as the provision of basic infrastructure for technological innovation will remain a major concern for most developing countries in the next decade.

The more advanced developing countries such as Brazil, South Korea, India and other newly-industrialized countries are more likely than the poorer countries to utilize the environmentally sound technologies available in the public domain.[13] In most African countries 80–90% of the total recurrent budgets of national R&D institutions is devoted to personnel emoluments.[14] Technological transition towards sustainable development is therefore likely to be uneven and will be largely influenced by the existing technological capacity in these countries.[15]

As research and science push the frontiers of knowledge, it is important to bear in mind that such efforts must not lose sight of the guiding principles of sustainable development. Indeed, "[s]cience – like art – is a listening-post at the outer edges of human perception. But science cannot work in isolation. For science to make maximum impact on the societies of tomorrow, it must interact with politics, with democratic debate, and it must be geared towards defined goals."[16]

The current efforts to integrate environmental considerations into all spheres of human endeavour will also influence the character of environmental research. It is likely to fade away as an explicit sectoral activity as other research programmes incorporate environmental objectives into their work. Other fields such as science and technology policy will also have to take on the challenges of sustainable development.[17] So far, this field has been slow to respond to the challenges, a sign of the slow rate at which institutions have responded to the issue.[18]

In addition to technological innovation, the pervasiveness of information technology is facilitating the diffusion of knowledge on environmental management. The growth in information networks and databases will assist greatly in the promotion of ideas on sustainable development. Countries that do not have easy access to such facilities are likely to be "isolated" from the development and to face difficulties in implementing sustainable development policies.

Ecotechnology Assessment

In a recent study, Vernon states that developing countries "that develop a strong internal capacity to search out and evaluate foreign technologies are usually able to acquire the technologies they need on satisfactory terms. Those that fail to search and evaluate, however, can make costly errors."[19] Technology assessment has emerged as an important discipline whose aim is to evaluate the prospects and risks associated with particular technologies.[20]

Technology assessment is one of the areas that African countries have paid the least attention to. The tendency has been to rely on technological information from those supplying the technology. In many cases, the ability of the African countries to assess technologies is prejudiced by tied aid and linkages between the supplies of technology and those providing finance (either as grants, equity capital or loans).

Technology assessment as a generic concept has now become part of the administrative language of many institutions in the industrialized nations and a few developing countries. The resurgence of the environmental agenda in the 1980s and the growth of analytical techniques for technology assessment has led to renewed emphasis on what is now called "environmentally sound technology assessment" or "ecotechnology assessment." *Our Common Future* recognizes the dual nature of radical technological change:

> Technology will continue to change the social, cultural, and economic fabric of nations and the world community. With careful management, new and emerging technologies offer enormous opportunities for raising productivity and living standards, for improving health, and for conserving the natural resource base. Many will also bring new hazards, requiring an improved capacity for risk management.[21]

The first reason for technology assessment is to provide a basis for long-term development policy. A second reason lies in the complexity of modern socio-economic systems. So rapid and unpredictable are the structural shifts, that merely stressing economic policy may be misleading. Much of the modern theoretical literature concentrates on the details of static assumptions about costs and prices and tends to ignore dynamic aspects of ecological change, innovation, development and linkages in the wider socio-cultural setting. Here, technology assessment is beginning to play a more important role.

The idea of technology assessment was first put forward in the US in the late 1960s. It resulted from a growing awareness of the adverse effects of technological development and was "devised to guide and control the development of new, large-scale, complex and extremely expensive technology in conformity with social goals." The focus then was based on the view that technological change was the main source of ecological problems.

The case for the transfer and acquisition of environmentally sound technology has thus become a major theme in international relations, as illustrated by the case of negotiations over the conventions on climatic change and biological diversity as well as Agenda 21. Given the difficulties of determining the ability of the developing countries to utilize the available technologies, it is necessary to establish mechanisms at the international and national levels that undertake technology assessment with a focus on environmental criteria.

Incremental Technical Innovation

Much of the discussion on technology transfer has focused on the importation of new technologies. This has led to arguments that place excessive emphasis on issues such as intellectual property protection and finances. While we recognize the importance of these issues, it is vital to stress the value of incremental technological innovations, especially at the firm level. Studies on industry in the newly-industrialized countries show clearly that much of what constitutes industrial dynamism is a result of the cumulative technological innovations introduced at the firm level.[22]

Such innovations are critical in the field of energy conservation where retrofitting and adjustments in energy use are important and can have more immediate savings than the installation of new equipment.[23] The variations in the intensity of energy use indicates where possibilities for incremental technical innovations to improve energy efficiency may lie. Over the 1970–90 period, energy intensity (or energy use per unit of GDP) has declined by 29% in the industrialized countries while it rose by 30% in the developing countries.

Such innovations need to be distinguished from "radical" or "major" innovations which are often associated with the introduction of new capital goods. Incremental technical innovations are introduced during the regular operations of plants and their cumulative effect over time may be more important than introducing new equipment.

Such innovations could play an important role, not only in improving energy efficiency, but also in re-designing plants so that they use alternative raw materials or reduce emissions of harmful gases. Incremental technical innovations are often associated with capacity-building at the firm level and emerge from training programmes as well as organizational change. On the whole, incremental technical change results from conscious policy efforts to enhance the capacity of workers to improve their performance, as well as the efficiency and output of equipment.

Environmentally Sound Technologies

International Considerations

The role of technology in environmental management has been gaining in currency over the last two decades.[24] Such recognition is also reflected in international conventions, such as the 1979 Long Range Transbounding Air Pollution Convention dealing with acid precipitation in Europe, the 1987 Montreal Protocol on substances that deplete the ozone layer (especially the 1990 London Amendment to the Protocol), the 1989 Basel Convention on the transboundary movements of hazardous wastes and their disposal, and more recently the 1992 Framework Convention on Climate Change as well as the Convention on Biological Diversity. Technological change concerns have now become a standard feature in international environment agreements.

Chapter 34 of Agenda 21 sees environmentally sound technologies in the context of pollution as process and product technologies "that generate low or no waste for the prevention of pollution." They also cover 'end of the pipe' technologies for treatment of pollution after it has been generated. They:

> are not just individual technologies, but total systems which include know-how, procedures, goods and services, and equipment as well as organizational and managerial procedures. This implies that when discussing transfer of technologies, the human resource development and local capacity-building aspects of technology choices, including gender-relevant aspects, should also be addressed. Environmentally sound technologies should be compatible with nationally determined socio-economic, cultural, and environmental priorities.

Environmentally sound technologies could fall into three main functional categories. The first category would include processes and materials which are developed for neutralizing or reducing the harmful effects of a given operation on the environment without necessarily introducing fundamental changes in the original process. The second category covers process modifications including the use of novel monitoring and control techniques, and changes in the raw or intermediate materials, which may be incorporated into existing technologies to eliminate or reduce their negative environmental impacts. The third category includes novel and traditional technologies which are inherently sound from the environmental point of view.[25]

Technology is the link between people and the resource base. The use of environmentally sound technologies therefore, implies a new evaluation of this relationship by re-assessing the methods used for the transformation of natural resources into useful products.[26] From the 1970s there has been a marked response from industry. New technologies and processes designed to reduce pollution and other environmental impacts have been developed.

Advances in biotechnology, sound energy technologies and alternatives to chlorofluorocarbons (CFCs) are notable examples of the response. However, these technologies are largely available in the industrialized countries which have the capacity to invest in their production, the institutional means to utilize them effectively and the means to monitor and mitigate environmental damage.

The level of technological development in the African countries is low; they are consequently more prone to technological dependence. The low rate of technological change may imply a sluggish transition towards the use of environmentally sound technologies. Further problems are created by the declining import capacity which began during the 1980s due to economic crisis. The decline of foreign direct investment during the same period has also weakened the ability of African countries to acquire imported technologies.

Environmentally Sound Technology Transfer

Since environmentally sound technologies are an integral part of policies for sustainable development and the dimension of environmental issues is global, there is a need for concerted action by nations. The international community is faced with the challenge of developing legal and trading regimes which promote the transfer of environmentally sound technologies to developing countries.

The categories of transferable technologies include capital goods, services and design specifications; skills and knowledge for production; knowledge and expertise for generating and managing technical change.[27] Technology transfer is not only the introduction of technology, but it also involves the imparting of the necessary knowledge and skills for the continual management of such technology. It is a two prolonged process, and for it to be effective, a strategy aimed at dealing simultaneously on inflows of foreign technology and on the development of local technological capacity is imperative. The process starts with developing the capacity to make appropriate choices based on technology assessment.

Apart from financial constraints, international trade regulations, patent laws and licensing regulations have been identified as some of the barriers to the transfer of environmentally sound technologies. However, it has been argued that:

relaxation of intellectual property and licensing restrictions will not necessarily lead to greater technology transfer. The failure of developing countries to use fully technological information in the public domain e.g.
patents that expired illustrates the problem. The ability to assimilate technology – not barriers to transfer is the primary impediment to technological development in developing countries.[28]

The ability to assimilate technology can be enhanced by capacity
building which embraces "the development of individual, group and
institutional capacity of self-sustained learning, generation of technology and implementation of developmental or scientific activities of a
defined range of problems."[29]

The concept of endogenous capacity in science and technology has
been described as the decisive pre-requisite in the management of technological change towards sustainable development. the potential of
environmentally sound technologies is said to be limited in effectiveness
unless that capacity is built. The primary functions of endogenous capability are enabling the efficient use of imported technology and the creation of technology with appropriate characteristics. Capacity therefore
encompasses the ability to: (a) make informed judgements on science
and technology matters; (b) select and utilize technologies; (c) adapt and
generate technologies; and (d) create new technologies. The prevailing
policy environment and incentive regimes play an important role in
technological capacity-building and utilization.

Technology and Climate Change Abatement

The burning of fossil fuels, loss of forests through logging and agricultural development coupled with rapid population growth, increase
the anthropogenic emissions of greenhouse gases into the atmosphere.
Carbon dioxide and other gases, including methane and CFCs, play a
major role in the "greenhouse effect". Nearly 56% of these emissions
come from energy generation through the burning of fossil fuels; about
21% from agricultural activities (paddies and cattle breeding) and
approximately 15% from CFCs. The concentration of carbon dioxide in
the atmosphere is also influenced by water pollution and deforestation
which causes a decrease in the natural cycle of carbon dioxide absorption.

Nations have at least two options: the first option is to take preventive
measures now in order to slow down the rate of greenhouse warming;
the second option is to undertake adaptive measures later to reduce the
impacts of climate change. Since predictions on the magnitude of
impacts are not specific, it makes sense to apply the precautionary principle and adopt the first option. This approach is reflected in the

Framework Convention on Climate Change which recognizes the need for countries to take immediate action as a first step towards comprehensive strategies on climate change.

Article 2 states the objectives of the Framework Convention on Climate Change as the stabilization of greenhouse gas concentrations in the atmosphere at a level that would prevent dangerous anthropogenic interference with the climate system. Such a level should be achieved within a time frame sufficient to allow ecosystems to adapt naturally to climate change, to ensure that food production is not threatened and to enable economic development to proceed in a sustainable manner.

Advances in technology show that climatic change abatement can be realized by the use of environmentally sound technologies. The technical potential to cut emissions of green house gases especially in the energy sector is significant. By using existing cost-effective technology, global emissions could be cut by 20% by the year 2020.[30] Such technologies can be used to: increase the efficiency of energy production and use; switch from carbon to hydrogen-based fuels; use carbon-free energy sources; reduce CFC emissions; and generally reduce rates of deforestation (which will in turn lead to expansion of carbon sinks).

The vital role of technology in climate change abatement is recognized by Article 4(c) of the Framework Convention on Climate Change whereby parties undertake to:

> promote and cooperate in the development, application and diffusion, including transfer of technologies, practices and processes that control, reduce or prevent anthropogenic emissions of greenhouse gases not controlled by the Montreal Protocol in all relevant sectors, including the energy, transport, industry, agriculture forestry and waste management sectors.

Article 4(e) strongly emphasizes the importance of technology in climate change abatement and urges the developed countries to:

> take all practicable steps to promote, facilitate and finance, as appropriate, the transfer of, or access to, environmentally sound technologies and know-how to other Parties, particularly developing country Parties, to enable them to implement the provisions of the Convention. In this process, the developed country Parties shall support the development and enhancement of endogenous capacities and technologies of developing country Parties. Other Parties and organizations in a position to do so may assist in facilitiating the transfer of such technologies.

The success of any programs aimed at strengthening the technological capacity of the developing countries will depend to a large measure on existing policies and incentive systems for technological development in particular, and innovation in general.

Technology Policy and Incentive Systems

Technology Policy and Sustainable Development

Developing countries have generally relied on the exports of primary goods and labour intensive manufactures based on natural resource endowment and availability of labour. It is now accepted that these countries need to enhance growth; they need to increase exports with higher technology value added by diversifying their products and markets. To achieve this, the technological base of the production structure needs to be strengthened. A sound technological base is indispensable in expanding the productive capabilities of developing countries. Further, technological innovations offer opportunities to fulfill their obligations in the goal for sustainable development.

Institutional and government policy can hamper or accelerate efforts to acquire and diffuse technologies. There is need for government adopted policies to induce rapid improvement in this area in view of the role of technology in development.[31] The role of national technology policies should be to ensure that the objectives of sustainable development and environmental protection are given special attention. To be effective, such policies must be comprehensive by addressing all functional sectors in society. They must also take into account socio-economic and cultural factors. National commitment to the objectives of such a policy is a necessary pre-requisite.[32]

Further, ideal science and technology policies are those which affect technological development directly by stimulating R&D, setting up a scientific infrastructure and giving preference to the output of indigenous technology. Regulation of access to foreign technologies is also important. Specific policy measures may include, firstly, larger allocations of national resources to R&D institutions. This can be fortified by other financial incentives for R&D activities. Secondly, they may include the establishment of suitable infrastructure to cater to important matters such as training, joint research programs, information availability and exchange and appropriate legal machinery for enforcement. Thirdly, the encouragement of acquisition and diffusion of encompasses which are appropriate to local conditions and resources are also included.[33]

Over the last three decades, there has been a marked decline in returns from the export of raw materials of which biological resources constitute a major part. Increases in production and utilization efficiency have continued to undermine the market for raw materials. Biotechnology itself is threatening a wide range of raw materials from the developing countries.[34] Some of these countries have placed emphasis on downstream processing as a way of adding value to their exports.

Such programmes have led to extensive importation of machinery and contributed to local industrial production and employment. Very few countries, however, have placed the local processing of raw materials in the broader context of technological innovation. Where plants have been installed, they have added little to the national technological capacity. It was often assumed that the mere installation of machinery would lead to the accumulation of technological competence. Indeed, much of the technological development in the African countries has been associated with the acquisition of equipment and machinery for the extraction of raw materials.

Although the policy instruments applied in various countries to promote technological innovation may look alike, it is important to recognize the underlying differences, especially in conceptions about international trade which is closely related to prospecting. The wide array of policy instruments used to promote technological innovation have received little attention in most developing countries.

In the Western world technology policy has often been conceived in the context of causal relationships. This is evidenced by the US approach whereby focus is placed on basic science, health, energy, agriculture and defence. The strategy adopted by Japan and the newly-industrialising countries (NICs) of the Pacific Rim, on the other hand, has tended to take a systems approach with a strong bias for industrial R&D. In the latter category of states research and commercial activities are closely focused on local and regional markets. These are important considerations when examining the transferability and relevance of some of these policy measures.

National policies on technology have often been influenced by differences in scientific and technological capability, industrial potential, as well as political and economic ideology. Technology policy measures started receiving increased emphasis in the industrialized market-economy countries in the 1960s. This followed the recognition that science and technology played a key role in the growth of the US economy over the previous 50 years. It was felt at the time, in the spirit of the Keynesian tradition, that government intervention was necessary to facilitate the role of science and technology in sustainable development.

It was also felt at the time that market and institutional imperfections affected the rate and direction of investment in R&D and government intervention was necessary. Researchers showed that firms were failing to make full use of their own research results. This feature is mainly seen in

TABLE 2.1: Government Innovation Policy Measures

Measure	Examples
Procurement	Central and local government purchases and contracts, public corporations, R&D contracts, prototype purchases, setting of design and performance criteria, choice of priority technologies
International trade	Trade agreements, technology acquisition agreements, tariffs, foreign exchange regulations, export compensation, import subsidies, licensing, infant industry protection, negotiation
Public enterprise	Innovation by publicly-owned industries, setting up new industries, pioneering use of new techniques by public corporations, participating in private enterprises
Scientific and technical	Research laboratories, support of research associations, learned societies, professional associations, research grants, setting up science parks
Education	General education, universities, technical education, retraining, vocational education
Information	Information networks and centres, libraries, advisory and consultancy services, databases, technology monitoring, liaison services, public awareness
Financial	Grants, loans, subsidies, financial sharing arrangements, venture capital, provision of equipment, buildings or services, loan guarantees, tariff remissions, export credits
Taxation	Company, personal, indirect and payroll taxation, tax allowances, depreciation allowances, tax exemption for private foundations
Legal and regulatory	Patents, utility models, plant breeders' rights, environmental and health regulations, contractual arrangements, conventions, inspectorates, monopoly regulations
Political	Planning, regional policies, honours or awards for innovation, encouragement of mergers or consortia, public consultation, creation of new institutions, setting up of research funds, initiating legal reforms
Public services	Purchases, maintenance, supervision and innovation in health service, public building, construction, transport, telecommunications, infrastructure, administrative guidance
External assistance	External aid, technical assistance, training, information provision
International relations	Sales organizations, trade and diplomatic missions (science attachés), technical co-operation, research representatives, international negotiations

Source: African Centre for Technology Studies, Nairobi.

circumstances where scientific knowledge is not properly integrated into the operations of the organisation, probably due to a failure to embody the research findings into applicable technology.[35]

In addition, risk, uncertainty and the high costs of R&D investment reduced the rate of investment in R&D. This situation, it was argued, would reduce the social benefits of R&D and shift research towards short-term goals. Policy intervention was therefore required to maximize the social gains of R&D.

There are at least four broad technology policy approaches adopted by the industrialized countries. A number of these countries have integrated technology policy into broad national development strategies. Such countries include Japan, France and Italy. In countries such as Germany, Denmark, and the Netherlands, technology policy is one of those policy instruments aimed at creating a suitable environment for economic change and international competitiveness. Technology policies in these countries are not formulated in the context of national plans.

This, however, may change as countries such as the Netherlands start to review the role of science and technology in sustainable development. In these countries, technology policy is used to shape the direction of economic change. These countries have set up a wide range of consultative and co-ordinating procedures as well as institutions within government and industry. In addition, other institutional measures have been introduced to facilitate the linkages between industry and government. What remains conspicuously lacking, however, is the issue of making the industrial policies sensitive to the ecosystem.

The Japanese model is the first category. The role of the Ministry of International Trade and Industry (MITI) in Japan has been a major subject of study.[36] What is significant is that MITI was conceived as an instrument for promoting technological innovation as a tool for international competitiveness, and not simply to promote trade in the context of comparative advantage.[37] The approach has been systemic from the beginning. The people who shaped MITI had no sympathy for conventional economic theory. Their aim was to promote the most advanced technologies with the widest possible world market.

The US has also introduced a wide range of measures aimed at enhancing its international competitiveness. Policy measures in the past have been selective, although government funding for new technologies has been a long-standing practice.

The second category includes a set of measures that allow governments to procure technologies at the early stages of their development. This approach was extensively used in the development of the electronics industry. In the US, for example, public procurement has been used to stimulate military technology. Other countries, especially in Europe, used public procurement to promote innovation in renewable energy technology in the 1980s.

The third category of technology policy measures covers financial support for innovation. Tax credits have been used in countries such as Japan, Canada and the US to stimulate and promote innovation. Other well-established measures such as grants, risk-sharing investment and loans are used in most industrialized countries. In addition, new financial support schemes are currently being introduced in the industrialized countries to support small and medium-size firms.

These three major categories are currently being supplemented by measures which focus on specific sectors and are implemented through international collaborative programmes. The European Research Coordination Agency (EUREKA), launched in 1985, was designed to bring together industry, university and government researchers in market-oriented information technologies in a bid to compete against the US and Japan. Other collaborative research measures include the Programme for Basic Research in Industrial Technologies for Europe (BRITE) and the European Strategic Programme of Research and Development in Information Technologies (ESPIRIT). In 1985, the European Economic Community (EEC) Council of Ministers approved a plan for a European Technological Community.

Unlike their industrial counterparts, the developing countries' technology policies have emphasized technology acquisition. These states have tied the transfer and acquisition of technology to negotiations regarding exploration of and access to biodiversity. A few developing countries, however, have introduced fiscal incentives and financial assistance. Countries such as South Korea, Singapore, Malaysia, Mexico and Peru have tax incentives.

Financial assistance, however, is more widespread in the developing countries. Singapore, for example, has experimented with a number of R&D and product development schemes in the last 10 years. Most developing countries are yet to come up with effective policies for technological development. Where such policies exist, their administration and management have been poor.

Policy and Incentives in Africa

Policy Environment. Given the current state of the African countries there is a need to identify a few systemic technology policy measures which can be used to stimulate and promote the sustainable development of indigenous technological capability. Such measures are defined as a policy interventions which have the capacity to achieve systemic gains by reorganizing the economic system as well as the institutional terrain with minimal investment, administrative requirements, staff and infrastructure. The main factors behind this approach is information flow, technical content and institutional networking. The approach differs from other policy formulation strategies in the sense that it utilizes the synergistic links between sectors. It uses a systems approach instead of relying on linear causal relationships.

Technology policy formulation is closely linked to prevalent trends in international trends and the related institutional arrangements. The current emphasis on technology as a tool for international competitiveness makes it increasingly difficult for most African countries to acquire emerging technologies. Although some of the experiences of other countries may be relevant for Africa, it is important to place these in the context of emerging design trends and institutional arrangements relating to the access to the technology and related information. Technology policy formulation has in the past been characterized by the enactment or publication of distinct laws or policy papers. The situation is changing and technology policy formulation is becoming a dynamic process which is guided by continuous review, analysis and research conducted by a wide range of institutions. The process requires continuous research and monitoring of both internal and international trends in innovation.

It has been stressed that achieving sustainable development will require major policy changes in the African countries. Such statements are often not backed by an appeal for studies which indicate the kinds of changes that need to be introduced. The tendency has been for the industrialized countries to suggest the extension of their policy measures to the African countries. This has often been possible because the discipline of policy research is still nascent in developing countries and virtually absent in Africa. Pronouncements of a few people who are familiar with the policy needs for specific changes have often been confused with policy research, which is a discipline in its own right. This problem is compounded by the fact that scientists have often believed that they should speak to power directly and that their research results form the rational basis for policy making.

This view, however, ignores the fact that interaction between decision-making and reality is not always rational and depends on a wide range of factors which are the subject matter for policy analysts. The area of policy research offers new possibilities for partnership between the industrialized countries and the African countries. These partnerships will essentially be research-oriented although they will also take on the role of information dissemination. The partnerships that are developed for research and information sharing will also serve the purposes of policy research.

One of the most important aspects of long-term social transformation is the rise and retention of institutions. Institution-building and reform is a good indicator of the rate at which any society is changing. The success or failure of facilitating sustainable development in Africa will depend to a large extent on the ability of existing institutions to implement the agents. So far, the view of African institutions by donor agencies have been limited to state institutions. While it is important to strengthen state institutions, it is necessary to broaden the scope of support to other institutions.

In the past, institution-building has been associated with providing infrastructure, finance and training. These three have often been treated separately and donors have tended to emphasize one at a time. It has become obvious in recent years that the provision of infrastructure and funding to institutions does not necessarily lead to increased capability to deal with development problems. In many cases, the funding has either been wasted or misused. There is therefore a need to re-think the nature of institution-building relevant to the African situation.

Institutions are arrangements through which people manage change. This implies that institutions are essentially about people. The ability of the people to manage change will depend largely on the level of their competence. Alternate ways of supporting institution-building in Africa should therefore be based essentially on competence enhancement. This view, which departs from the narrow approach of providing funds and infrastructure, opens up a wide range of opportunities for designing new partnerships in institution-building. It also widens the range of institutions that can be involved in the process.

So far, training programmes have been too restrictive and often designed in the context of academic specification. There may be a need to re-think the orientation of the academic specifications so as to provide a suitable environment for developing competence in environmental management. The conventional view has been to think of training in the context of where it takes place. As a result, the criteria for training programmes tends to restrict the choice of either local or foreign institutions. While it is important to build competence in specific African countries or regions, it is also important to create conditions under which training can be used as a means of transferring competence from the industrialized countries to Africa.

Africa is marked by a wide range of institutional arrangements aimed at formulating science and technology policies. These arrangements include ministries or ministerial committees, science planning bodies, multi-sector coordinating bodies and scientific research coordinating bodies. Some of these institutions have statutory authority while others do not. Over the years, science and technology policy-making has been moving from councils and committees to ministerial organs.

TABLE 2.2: Trends in Science and Technology Policy Institutions, 1973–1986

	1973	1979		After 1979	Total 1986
		Increase	Total		
Ministry of Science or Ministerial Science Policy Council	5	+4	9	+18*	27
Science planning body in general	12	+16	28	+2	30
Multisector body for coordinating scientific research	18	+6	24	+4	28
Natural science research	2	+14*	16	+9	25
Agricultural research	15	+15*	30	+2	32
Medical research	6	+14	20	+1	21
Nuclear research	3	+1	4	-	4
Industrial research	7	+15*	22	+3	25
Environmental research	2	+13	15	1	16
Total	70	98	168	40	
Grand total			168	40	208

* Indicates bodies with rapid increase.

Source: Forje, J. 1987.

Many of these institutions were established after the United Nations Conference on Science and Technology for Development (UNCSTD) held in 1979 in Vienna which recommended that every country should formulate a national science and technology policy. Table 2.2 shows the status of the evolution of these institutional arrangements in Africa.

Incentive Schemes. In most African countries, incentives for technological development have often been implicit and mediated through investment incentives for industry. Where explicit incentives have been provided for technological development, they have often remained on the books or only articulated in policy documents.[38] What emerges so far is that the main incentives for industrial development have favoured the importation of large-scale, capital-intensive investments with little consideration for technological development.[39] In many cases, the incentives explicitly provided for industry have tended to implicitly undermine the prospects for technological development.

For example, incentives provided in many countries to facilitate the installation of industrial plants have often worked against local technological development. These incentives often provide for lower import duties and tariffs on whole plants and impose higher import duties and tariffs on raw materials that could be used to fabricate such equipment locally. The effect has often been that equipment that could be fabricated locally is often imported.

Many African countries have the potential to manufacture most of the components necessary for assembling solar panels. But the high tariffs imposed on the necessary raw materials makes it easier for firms to import complete sets of panels. But because of the high labour costs in the industrialized countries, such products are usually expensive and only a few institutions and a small section of the population are therefore able to secure solar panels. Relaxation of tariffs would make it possible for cheaper solar energy panels to be manufactured in some African countries.

In many cases, tariff regulations are simply insensitive to the imperatives of technological development. For example, despite the growing importance of computer equipment in scientific research, many countries still place high tariffs on such equipment. Part of the problem lies in the misconception propagated in the 1970s that computerization was likely to displace jobs. Those countries which had strong job-creation policies imposed high tariffs on computer equipment.

Commercial organizations and individuals may not find it worthwhile to take into account environmental factors in the technologies which they develop or use. Entrepreneurs will be motivated by the existence of a suitable market before developing technologies. Incentives are therefore important in that they create conditions which encourage involvement in a new area. Incentives cannot be established unless there is a clear government policy on technology and development.

After independence most African countries established tariffs and import licensing systems to encourage investment in import substitution industries. The industries depended heavily on imported machinery and inputs. Some of these countries also liberalized foreign investment policies by taking measures to protect foreign investors against nationalization, reduce limitations on share ownership, increase field of activities and size of profit remittances and accept arbitration as a means of settling disputes. These incentives proved useful in guiding activities of investors into areas which were deemed to be priorities. However, these incentives had their own limitations.

Experience has shown that the most effective and efficient inducements are those arising from financial stability, policy transparency, availability of skilled manpower and large growing domestic markets. More incentives are needed for those investments which encourage indigenous capacity building. Incentives are also needed in areas that provide inputs which firms and individuals are unable or unwilling to provide themselves or which help to reduce the risk of uncertainty associated with certain types of investment in the accumulation of knowledge and skills.

Manufactures in most African countries receive no income tax concessions, tariff reductions or other incentives for their R&D efforts. Also, most research activities are not well coordinated, despite the establish-ment of national institutions for the purpose.

Most intellectual property laws in Africa do not encourage the use of knowledge which is already in the public domain. These laws are also deficient in that there is no room for appropriate technologies developed by indigenous people.[40]

The barriers encountered by transnational corporations in the transfer of environmentally sound technologies include: (a) reduced expectations of profits from sales of such technologies due to lack of markets, com-plex legal requirements, lack of market information; (b) lack of adequate technical and social infrastructure; and (c) unfair competition due to lack of environmental regulations and standards in developing countries.

During the 1980s, many developing countries relaxed regulations governing technology and foreign investment in their countries. Policy changes easing controls over patents, licensing and trademarks were introduced. These efforts were a recognition of the importance of tech-nology and also the importance of international collaboration in this area. However, more direct incentives need to be provided to support technological development in general and environmentally sound tech-nologies in particular.

Strengthening Incentive Systems

Overall Policy Environment

The growing liberalization of economic systems worldwide has changed the role of the state in development and altered the patterns of policy making. Whereas African governments have tended to invest directly in enterprise development, their role in the 1990s will become more regulatory. This, however, does not mean that the state will play a lesser role in development activities. To the contrary, there is a need for the state to provide the necessary incentives and improved policy envi-ronment for entrepreneurial activities. In most countries the policy envi-ronment is still hostile to private enterprise activities. The challenge for governments is not how to recede into the background, but how to improve the policy environment for entrepreneurial development.

In some sectors of the economy, especially in technological develop-ment, traditional policy measures may be necessary even though the general trend is to use other economic instruments. The issue is not so much the use of interventionist measures, but the discipline that is need-ed in applying such measures. Korea gives an example of discipline in the use of subsidies to promote technological innovation.[41]

The need for discipline and long-term considerations in the use of instruments such as subsidies are becoming more critical as new evidence on the rate of industrial learning in infant industries emerges. "Evidence from . . . Korea, suggests that the learning time in the engineering industry is much longer than anticipated; two decades seems not unusual."[42] Further, it "is also argued that due to both the speed of technical change and the increased globalisation of industries, the learning time has extended in the past 15 years and the social cost of fostering the infant industries has increased."[43]

Before one considers the specific incentives that can be used to promote the application of environmentally sound technologies, it is important to consider the overall policy environment that these incentives operate. The issue of time is essential to the process. Since technological development requires long-term planning, the policy environment must be predictable. In this respect, policy instruments should not be changed without considering their long-term impacts on technological investment. Provisions need to be made to ensure that those who are affected by policy changes are adequately compensated.

Compensation, however, should not be used as an excuse to introduce unexpected changes in policy instruments. Where the policy environment is predictable and reliable, entrepreneurs are more likely to adapt to changing market conditions. For example, a country that puts in place clear long-term goals to select technologies on the basis of their environmental soundness will provide signals to entrepreneurs to start investing in meeting the demand for such technologies.

In addition to predictability, issues such as participatory policy-making are becoming equally critical to the development of technologies. In most African countries, policy-making is treated as a secret activity performed only by certain individuals in a few state agencies. This non-participatory practice makes it difficult for entrepreneurs to make long-term plans or to contribute ideas to the policy-makers. A participatory policy-making process is inherently transparent and brings the stakeholders into partnerships with governmental and non-governmental agencies in ways that facilitate the development process.

Trade Policies

Trade policies are important determinants of the prospects for successful investment in environmentally sound technologies. In the past, trade policies have relied on protectionist measures. However, most African countries are now shifting towards export-directed trade policies. Trade in environmentally sound products is going to become an important aspect of trade in general. In this regard, countries may formulate trade policies that specifically promote trade in "green products".

While protecting infant industries is deemed to create unnecessary inefficiencies, it may be necessary to provide learning-related protection in certain areas of industrial activities. It is unlikely that African countries will accumulate the necessary technological capacity unless their investments are partially shielded from direct competition with the more industrialized countries. But such protection must be directly linked to industrial learning, and not seen as a way of protecting inefficient production methods. Industrial learning would manifest itself in increased production efficiency and accumulation of technological capacity at the national level.

Market-Oriented Procurement

The role of public procurement in promoting technological innovation has not been given adequate attention in the African countries, especially given the lack of measures such as venture capital which could be used to promote technological development. So far, most African countries operate tender systems which emphasize comparative prices. Under such a system, the lowest tenders are usually awarded contracts. This helps to reduce public expenditure.

However, it also undermines the capacity of local manufacturers to enter the market, especially where the products are new and the prices are high due to low volumes. A public procurement policy would ensure that local products are sold and that markets are created through government purchases.

Since government is still a major economic actor in Africa, a procurement policy would enable the government to put pressure on manufacturers to improve the quality of their products, reduce prices, maintain certain performance standards and improve the design. The difference in cost between the lowest price of a product and the cost of the local product could be considered as government investment in R&D. With this level of investment, the government would have the power to influence and direct the pace of technological innovation.

Implementing a technology-based procurement policy would require a change in the composition of skills in the tendering agency. This would require engineers, designers, materials scientists and other relevant technologists depending on the priority products supported by the government. The policy would also require a review of the current standards policy adopted by most African countries. So far, standards are enforced without due consideration to the technological needs of the country. They normally do not account for the fact that standards evolve and therefore efforts should be made to allow for this gradual improvement.

It should be pointed out that procurement programmes can also be abused. Since the procuring agencies will have extensive powers and influence on the direction of technological change, it is important that criteria for choice of technology be made in the context of national priorities. In addition, government procurement programmes should emphasize innovations that come from the private sector or those for which commercialization options already exist.

This will force public sector R&D institutions to forge links with the private sector. There is always the danger of the state procuring from itself in a such a way that the technology does not get established in the market place. The aim of the procurement programme should be to promote technological innovation in the pre-commercialization stages up to the stage when the technologies have established their own market niches. This can be achieved by setting market targets which the firm must achieve in order to continue receiving support through procurement programmes.

Intellectual Property Rights

One of the incentive regimes that has been widely covered in the literature is intellectual property protection. In the past, this regime has not provided particular attention to environmental protection but it could form an important area of the incentive system, especially at the stage of technology development and innovation.

Investment Codes

Creating incentives for environmentally sound technology development will require specific changes in the existing investment codes of various countries. The codes outline the benefits offered, criteria for eligibility and the obligations of investors as well as governments. Investment codes of most African countries provide fiscal and tariff concessions to enterprises that meet specified conditions which relate to the use of domestic inputs, firm size, employment creation, choice of site and sector, and others. The use of environmentally sound technologies could be included as part of the criteria for fiscal and tariff concessions.

Tariff Concessions and Disincentives. Tariff concessions could be used to promote investment in environmentally sound technologies. One way of doing this is to provide import duty exemption for raw materials necessary for the manufacture of environmentally sound products. Another way is to reduce duty for imported products that are environmentally sound in the general framework of trade liberalization. Where such concessions are made, additional measures would need to be introduced to ensure that local manufacturers who are likely to be affected by this provision are provided with support to enable them to improve the environmental standing of their products. It may be necessary for countries to establish national financial schemes that emulate the Global Environment Facility (GEF), which would assist in this transition.

Countries may wish to protect their natural resource base and human health through the introduction of higher tariffs for products that are known to be harmful, especially in cases where alternative technologies exist. Such measures should not be used as a way of restricting trade and should be carried out within the limits of the rules of the General Agreement on Tariffs and Trade (GATT).

Export Incentives. Many African countries are starting to offer export incentives which include exemption from export tax as well as export compensation. Other measures include tax rebates for duty paid on imported inputs, preferential tax on export earnings, retention of part of export earnings in foreign exchange, access to export processing zones, and export insurance. Such incentives could be used to promote the export of environmentally sound products and take advantage of the growing market for "green products" in the industrialized countries.

Tax Concessions and Disincentives. Tax concessions are the most commonly used incentive for technological development and can be used to support the development of environmentally sound technologies. So far, most African countries do not provide tax concessions for R&D expenditures. But such concessions, especially in operating plants, could stimulate incremental innovations that would reduce energy consumption or reduce pollution emissions. They could also encourage industrialists to switch to new raw materials or fuels for which they can receive concessions.

In addition to tax concessions, tax disincentives could also be employed to discourage the use of environmentally unsound technologies. The introduction of tax disincentives would only be feasible if environmentally sound technological options are available to the entrepreneurs and clear environmental standards have been introduced. In the absence of such options, industrialists are likely to object to the introduction of such disincentives.

Tax Holidays. A tax holiday, or the full or partial exemption from income and other taxes for a period, could be used to stimulate investment in environmentally sound technologies. This measure has been widely used and has in recent years been integrated into operation of export processing zones (EPZs). Many of the EPZs have more emphasis on export promotion than on technological development.

Accelerated Depreciation. Accelerated depreciation is a form of "tax holiday" which allows a firm to write off the cost of its capital equipment or scientific instruments against its gross revenue. This reduces the cost of capital and increases the liquidity of the firm. Accelerated depreciation would allow firms to invest in new facilities that are environmentally sound.

Investment and Reinvestment Allowances. Reinvestment allowances, which are used to encourage firms to expand, could be used to promote investment in pollution abatement at the firm level or incremental improvements that save on energy or raw material use. Such allowances exempt firms from tax on reinvested capital. This is a more positive measure than the practice of taxing savings or tax-exempt corporate profits that are distributed to the shareholders as a way of forcing firms to reinvest.

Profit Repatriation. Incentive schemes that allow for high percentages of profit repatriation could attract foreign investment. For such a scheme to work for environmentally sound technologies, it would have to give preferential treatment to this sub-sector on the basis of certain environmental standards.

Offsets

Offsets are economic benefits directed at national goods. They are essentially "off-balance sheet" and are not associated with direct costs to the government. They could encourage firms to use their capabilities to invest in the development of environmentally sound technologies and practices. The use of offsets could encourage entrepreneurs to give services such as management skills at no or reduced cost.

Infrastructure and Support Services

The development of infrastructure for the development of environmentally sound technologies is essential for long-term capacity building. The issue is of critical importance to Africa because of the deteroriation of infrastructure and declining ability of governments to support new investments. The decline in infrastructure may lead to the concentration of new technological investment near or in urban areas. The required infrastructure may range from the supply of critical equipment to the establishment of "science parks" or "industrial estates" and support to institutions of higher learning.

Access to information on environmentally sound technologies is a key service for industry and forms part of the incentive system. The institutions providing such services would need to ensure that systematic information on sources and prices of technology also includes advice and assistance in bargaining, details on trade restrictions and practices, warranties and the scope of proprietary rights.

Insurance and Reinsurance

African countries could involve insurance firms in covering damages that might occur during production, transportation, storage, utilization, and final disposal of environmentally sound products. This would assist in at least two ways: Firstly, assessment of risks and premiums that would encourage companies and institutions to develop and promote environmentally sound technologies. Secondly, the insurance company will assist in inspecting and regulating the means to improve environmental risk management.

Employment Incentives

In order to attract people into research-related jobs, some countries such as Kenya have created separate schemes of service researchers. Similar schemes could be designed to attract people into institutions that work on environmentally sound technologies. Other ways of creating employment incentives for environmentally sound technologies could be to provide tax deductions on activities, such as training, which are critical to capacity building.

Interest Rates

Many countries have used interest rates to stimulate borrowing for investment in certain sectors of the economy such as agriculture. Such measures could be applied for promoting environmentally sound technologies. But low interest rates could reduce the amount of capital available for lending, making it necessary to introduce interest rate subsidies. The effectiveness of this measure as a way of promoting environmentally sound technologies does not seem to be high.

Conclusion

The aim of this chapter was to assess the incentive regimes for the transfer and development of technological capacity in Africa with specific reference to environmentally sound technologies. It argues that much emphasis has been put on the international obstacles to the transfer of environmentally sound technologies, but the role of appropriate incentive regimes in the developing countries to facilitate technology acquisition and development has been downplayed. The chapter proposes a range of policy incentives that could be applied to promote environmentally sound technology development in Africa. It stresses that the effective application of the proposed incentives can only be achieved if implemented within a policy framework that puts technology at the heart of the sustainable development process.

Notes

1. This chapter is based on an on-going ACTS study, The Guiding Hand: Technology, Environment and Public Intervention, 1993.

2. Schumacher, E.F., 1973.

3. ACTS, 1991.

4. Ominde, S. and Juma, C. eds., 1991; Ottichilo, W. et al., eds, 1991.

5. This, however, should take into account the risks associated with urbanization and population migration.

6. This is exemplified by the endorsement of the International Chamber of Commerce's Business Charter for Sustainable Development (BCSD) by over 600 private corporations. The adoption of such a code of behaviour, if extended to their sub-contractors, could significantly contribute to the promotion of sustainable development.

7. For a detailed collection of case studies on this theme, see Schmidheiny, S., 1992.

8. The emerging trends suggest that for most developing countries, the most suitable starting point is the application of environmentally sound technology to small- and medium-scale enterprises, which are also the main sources of employment and also pose widespread environmental risks.

9. Brown, M., 1992, p. 12.

10. Stegmann, K., 1989, pp. 73–100. While governments continued to formulate policies based on national competition, firms increased their degree of cooperation, especially in the field of technology transfer. For details of such arrangements, see Juma, C. and Sagoff, M. 1992; Hagendoorn, J., 1990, pp. 17–30; Levy, J. and Samuels, R., 1991.

11. Brown, M., 1992, p. 13.

12. The establishment of Thailand's Environmental Research and Training Center with support from the Japanese government illustrates the point. See Tabucanon, M. 1991.

13. Marcovitch, J. et al., 1991.

14. See, for example, the data for Eastern and Southern African provided by Nyiira, Z.M., 1991.

15. The challenge for the international community is therefore to identify ways of enhancing the capacity of the developing countries to utilize the emerging and available environmentally sound technologies.

16. Brundtland, G.H., 1991, p. 9.

17. UNCTAD, 1990.

18. Aubert, J.-E., 1992, p. 5.

19. Vernon, R., 1989.

20. For a more detailed application of technology assessment to biotechnology, see Clark, N. and Juma, C., 1991.

21. World Commission on Environment and Development, 1987, p. 217.

22. Lall, S., 1987.

23. Bell, M., 1990; Chantromonsklasri, N., 1984.

24. Juma, C. and Ojwang, J.B. 1992.

25. For more details, see UNCSTD, 1991.

26. UNCTAD 1991.

27. Bell, M. 1990.

28. Juma, C. and Sagoff, M., 1991, p. 272.

29. Bekoe, D.A. and Prage, L., 1991.

30. Hanisch T. ed., 1991.

31. Hanisch, T. ed., 1991, p. 282.

32. Bell, M., 1990.

33. Bell, M., 1990.

34. Sasson, A., 1992.

35. Pavitt, K. 1987; Ghai, D., 1974.

36. See, for example, Johnson, C. 1982; Okimoto, D., 1989.

37. "The MITI decided to establish in Japan industries which require intensive employment of capital and technology, industries that in consideration of comparative cost of production should be the most inappropriate for Japan, industries such as steel, oil-refining, petro-chemicals, automobiles, aircraft, industrial machinery of all sorts, and electronics, including electornic computers. From a short-run, static viewpoint, encouragement of such industries would seem to conflict with economic rationalism. But from a long-range viewpoint, these are precisely the industries where income elasticity of demand is high, technological progress is rapid, and labour productivity rises fast," OECD, 1972.

38. See, for example, Goka, A.M. et al., 1990.

39. "The different types of incentives . . . can be categorized generally as special concessions made available under investment and tax codes, protection through trade and industrial licensing policies, credit and equity participation facilities through development finance corporations, and favourable interest rates. Provision of adequate infrastructure and services and creation of a stable environment were also seen as important means of attracting investment. Investment codes tended initially to be oriented toward large-scale foreign enterprises, with subsequent efforts to include domestic investors, especially in small-scale enterprises. The benefits generally depended heavily on the amount of capital invested or made capital available at low cost (both financial capital and imported capital goods). This had, however, the effect of biasing the structure of incentives in favor of large, capital-intensive investments," Steel, W. F. and Evans, J. W., 1984.

40. Juma C. and Sagoff M., 1992, p. 278

41. "In late-industrializing countries, the state intervenes with subsidies deliberately to distort relative prices in order to stimulate economic activity. This has been as true of Korea, Japan, and Taiwan as it has been in Brazil, India and Turkey. In Korea, Japan, and Taiwan, however, the state has exercised discipline over subsidy recipients. In exchange for subsidies, the state has imposed performance standards on private firms. Subsidies have not been giveaways, but instead have been dispensed on the principle of reciprocity. With more disciplined firms, subsidies and protection have been lower and more effective than otherwise," Amsden, A., 1989.

42. Jacobsson, S. Undated.

43. Jacobsson, S. Undated.

References

ACTS 1991. *Sustainable Industrial Development in Africa: Agenda for the 1990s.* Mimeo. Nairobi: African Centre for Technology Studies.

Amsden, A. 1989. *Asia's Next Giant: South Korea and Late Industrialization.* New York: Oxford University Press.

38

Aubert, J.-E. 1992. "What Evolution for Science and Technology Policies?" *The OECD Observer*, February/March.

Bekoe, D.A. and Prage, L. 1992. "Capacity Building". In Dooge, J.C. et al. eds. *An Agenda of Science for Environment and Development into the 21st Century*. Cambridge, UK: Cambridge University Press.

Bell, M. 1990. *Continuing Industrialisation, Climatic Change and International Technology Transfer*. Brighton, UK: Science Policy Research Unit, University of Sussex.

Brown, M. 1992. "Science, Technology and the Environment." *The OECD Observer*, February/March.

Brundtland, G.H. 1991. *Environmental Challenges of the 1990's: Our Responsibility Towards Future Generations*, Cambridge, UK: The Tanner Lecture on Human Values, February 14, Clare Hall.

Chantromonsklasri, N. 1984. *Technological Response to Rising Energy Prices: A Study of Technological Capacity and Technical Change Efforts in Energy-Intensive Manufacturing Industries in Thailand*. DPhil. Thesis, Brighton, UK, Science Policy Research Unit, University of Sussex.

Clark, N. and Juma, C. 1991. *Biotechnology for Sustainable Development: Policy Options for Developing Countries*. Nairobi: Acts Press, African Centre for Technology Studies.

Coughlin, P. 1991. "Industrial Development, Technological Change, and Institutional Organization: The Kenyan Case" in Nyong'o, P.A. and Coughlin, P. eds., *Industrialization at Bay: African Experiences*. Nairobi: African Academy of Sciences.

Coughlin, P. and Ikiara, G.K. eds. 1988. *Industrialization in Kenya: In Search of a Strategy*. Nairobi: Heinemann Kenya and London: James Currey.

Forje, J.W. 1987. *Trends in the Development of Science and Technology in Africa Since CASTAFRICA I*. Paris: United Nations Educational, Scientific and Cultural Organisation.

Fransman, M. and King, K., eds., 1984. *Technological Capability in the Third World*. London: Macmillan.

Galenson, A. 1984. *Investment Incentives for Industry: Some Guidelines for Developing Countries*. World Bank Staff Working Paper No. 669. Washington, DC: World Bank.

Ghai, D. 1974. *Social Science Research on Development and Research Institutes in Africa*. Discussion Paper No. 197. Nairobi: Institute for Development Studies, University of Nairobi.

Goka, A.M. et al. 1990. *Performance Review of Institutions for Technology Policy in Ghana, Nigeria and Tanzania*. IDRC Manuscript Report 241e. Ottawa: International Development Research Centre.

Hagendoorn, J. 1990. "Organisational Modes of Inter-firm Co-operation and Technology Transfer." *Technovation*, Vol. 10, No. 1, pp. 17–30.

Hanisch, T. ed. 1991. *A Comprehensive Approach to Climate Change*. Oslo: Center for International Climate and Energy Research-Oslo.

Heaton, G. et al. 1991. *Transforming Technology: An Agenda for Environmentally Sustainable Growth in the 21st Century*. Washington, DC: World Resources Institute.

Heaton, G. et al. 1992. *Backs to the Future: US Government Policy Toward Environmentally Critical Technology.* Washington, DC: World Resources Institute, Washington.

Jacobsson, S. Undated. *The Length of the Infant Industry Period: Evidence from the Korean Engineering Industry.* Department of Industrial Management and Economics. Mimeo. Göteborg, Sweden: Chalmers University.

Johnson, C. 1982. *MITI and the Japanese Miracle: The Growth of Industrial Policy, 1925–1975.* Stanford, California: Stanford University Press.

Juma, C. and Ojwang, J.B. eds. 1989. *Innovation and Sovereignty: The Patent Debate and African Development,* Nairobi: African Centre for Technology Studies.

Juma, C. and Ojwang, J.B. 1992. *Technology Transfer and Sustainable Development.* Ecopolicy No. 2. Nairobi: Acts Press, African Centre for Technology Studies.

Juma, C. and Sagoff, M. 1992. "Policies for Technology", in Dooge, J. C. et al. eds., *An Agenda of Science for Environment and Development into the 21st Century.* Cambridge, UK: Cambridge University Press.

Kemp, R. and Soete, L. 1990. "Inside the 'Green Box': On the Economics of Technological Change and the Environment." in Freeman, C. and Soete, L. eds., *New Explorations in the Economics of Technological Change.* London: Pinter Publishers.

Lall, S. 1987. *Learning to Industrialize: The Acquisition of Technological Capabilities by India.* London: Macmillan.

Levy, J.D. and Samuels, R.J. 1991. "Institutions and Innovation: Research Collaboration and Technology Strategy in Japan," in Mytelka, L. K. ed., *Strategic Partnerships and the World Economy: States, Firms and International Competition.* London: Pinter Publishers.

Marcovitch, J. et al. 1991. *Brazilian Experiences on Management of Environmental Issues: Some Findings on the Transfer of Technologies,* Research Paper No. 1., Geneva, United Nations Conference on Environment and Development.

Mkandawire, T. 1991. "De-industrialization in Africa," in Nyong'o, P.A. and Coughlin, P. eds. *Industrialization at Bay: African Experiences.* Nairobi: African Academy of Sciences.

Nyiira, Z.M. 1991. *Research Resources in National Research Institutions in Eastern and Southern Africa.* IDRC Manuscript Report 290e, Ottawa, International Development Research Centre.

O'Connor, D. 1991. *Policy and Entrepreneurial Responses to the Montreal Protocol: Some Evidence from the Dynamic Asian Countries.* OECD Development Centre Technical Paper No. 51. Paris, Organisation for Economic Co-operation and Development.

OECD 1972. *The Industrial Policy of Japan.* Paris: Organisation for Economic Co-operation and Development.

Okimoto, D. 1989. *Between MITI and the Market: Japanese Industrial Policy for Higher Technology.* Stanford, California: Stanford University Press.

Ominde, S.H. and Juma, C., eds. 1991. *A Change in the Weather: African Perspectives on Climatic Change.* Nairobi: Acts Press, African Centre for Technology Studies.

OTA 1992. *Green Products by Design: Choices for a Cleaner Environment.* Washington, DC: Office of Technology Assessment.

Ottichilo, W. et al., eds. 1991. *Weathering the Storm: Climatic Change and Investment Policy in Kenya*. Nairobi: Acts Press, African Centre for Technology Studies.

Pavitt, K. 1987. "The Objectives of Technology Policy." *Science and Public Policy*, Vol. 14 No. 4, pp. 182–188.

Rath, A. and Herbert-Copley, B. 1992. *Technology and the International Environmental Agenda: Lessons for UNCED and Beyond*. Ottawa: International Development Research Centre.

Sasson, A. 1992. *Biotechnology and Natural Products: Prospects for Commercial Production*. Nairobi: Acts Press, African Centre for Technology Studies.

Schmidheiny, S. 1992. *Changing Course: A Global Business Perspective on Development and the Environment*. London: MIT Press.

Schumacher, E.F. 1973. *Small is Beautiful: A Study of Economics as if People Mattered*. London: Blond and Briggs.

Steel, W.F. and Evans, J.W. 1984. *Industrialization in Sub-Saharan Africa: Strategies and Performance*. World Bank Technical Paper No. 25. Washington, DC, World Bank.

Tabucanon, M.S. 1991. *Thailand's Experience on Environmental Research and Training*. Research Paper No. 23. Geneva, United Nations Conference on Environment and Development.

Tiffin, S., Osotimehin, F. with Saunders, R. 1992. *New Technologies and Enterprise Development in Africa*. Paris: Organisation for Economic Co-operation and Development.

UNCED 1992. *Agenda 21*. Geneva: United Nations Conference on Environment and Development.

UNCTAD 1990. *Transfer and Development of Technology in Developing Countries: A Compendium of Policy Issues*. United Nations Conference on Trade and Development, New York, United Nations.

UNCTAD 1991. *Accelerating the Development Process: Challenges for National and International Policies in the 1990s*. Geneva: United Nations Conference on Trade and Development.

United Nations 1979. *Convention on Long-Range Transboundary Air Pollution*. Geneva: United Nations Economic Commission for Europe.

United Nations 1985. *Vienna Convention for the Protection of the Ozone Layer*. New York: United Nations.

United Nations 1987. *Montreal Protocol on Substances that Deplete the Ozone Layer*. New York: United Nations.

United Nations 1988. *Protocol to the 1979 Convention on Long-Range Transboundary Air Pollution Concerning the Control of Emissions of Nitrogen Oxides or their Transboundary Fluxes*. Geneva: Economic Commission for Europe, United Nations.

Vernon, R. 1989. *Technological Development: The Historical Experience*. Economic Development Institute Seminar Paper No. 39, Washington, DC, World Bank.

World Commission on Environment and Development 1987. *Our Common Future*. Oxford, UK: Oxford University Press.

3

National Greenhouse Gas Inventories and Other Information to Implement the Framework Convention on Climate Change

Jane A. Leggett

The Climate Convention established a framework for countries to take actions toward a common goal of stabilizing greenhouse gas concentrations in the atmosphere at a level that would prevent dangerous anthropogenic interference with the climate system. Whether that goal is ultimately met will depend on the good faith efforts of Parties to the Convention to undertake policies and measures which reduce their emission of greenhouse gases into the atmosphere and which enhance the uptake and permanent storage of carbon in reservoirs. The Framework Convention on Climate Change (FCCC) contains several provisions which, together, can facilitate achievement of the Convention's goal (UNGA 1992). Key provisions include:

- development and communication of national greenhouse gas inventories (Articles 4.1 and 12.1);
- detailed reporting of national policies and measures, and projections of their expected effects (Article 12.2);
- review by the Parties of the adequacy of commitments of Parties listed in Annex I (developed countries) on national policies and measures to mitigate climate change, including assessment by the Subsidiary Body for Implementation of the aggregated effects of Parties' policies and measures;
- cooperation to develop and exchange data related to the climate system and, inter alia, the causes and timing of climate change and the consequences of response strategies [Articles 4.1(g) and (h)].

The cumulative effectiveness of these mechanisms will be critical in motivating and assisting governments to fulfill their commitments under the Convention and subsequent protocols.

This chapter attempts to show how provisions regarding greenhouse gas inventories, national policies and measures,[1] and scientific cooperation could complement each other to enhance the effective implementation of the FCCC. The information provided through these three processes will allow the Conference of the Parties to determine the adequacy of the existing commitments. The information can be either an 'early warning' of a need to revise the current agreement, or assure policy-makers that they are on the right track. Hence, the mechanisms for developing and reviewing the required information will be important to the overall success of the Convention. A premise underlying the thoughts in this paper is expressed by a recent report of the U.S. Government Accounting Organization (GAO 1992). It concludes, "in a report examining eight major [international environmental] agreements, that they are not well monitored and, furthermore, that some parties to the agreements, especially developing countries, lack the ability to comply." There is good reason to believe that the climate convention is already better prepared than previous agreements to meet the challenge of implementation.

National Inventories of Greenhouse Gas Sources and Sinks

Articles 4.1(a) and 12.1(1) of the FCCC call on all countries to prepare and communicate national inventories of greenhouse gas sources and sinks, using comparable methodologies to be promoted and agreed upon by the Conference of the Parties. National inventories will be essential for governments to understand their contributions to climate change, to set priorities for actions to reduce sources and enhance sinks of greenhouse gases (GHG), and to measure the effectiveness of their policies and measures over time.

A large range of uncertainty surrounds current estimates of sources and sinks, although the range varies by country and source or sink. The good news is that a very useful process is already underway, under the leadership of the Intergovernmental Panel on Climate Change (IPCC), with support from the Organisation for Economic Cooperation and Development (OECD), to facilitate development of national inventories of greenhouse gas sources and sinks.

Since 1991, experts from over 44 countries have been:

- developing and recommending by mid-1993 guidelines for estimating sources and sinks of greenhouse gases[2] and their precursors (CO_2, CH_4, N_2O, NO_x, CO and non-methane VOC);

- training and assisting participants from developing countries to develop their own national inventories; and
- experimenting with "transparency studies," aimed at determining useful reporting and review procedures to assure that national inventories are reasonably comparable, reproducible and reliable.

While one should expect continual improvement in estimation methods, the IPCC will make adequate methodological guidelines available for the first meeting of the Parties to the FCCC. The estimation methods will represent the state-of-the-art for many categories, and will provide methods simple enough for countries to use even if they have little expertise or disaggregate data. A few categories of sources and sinks are likely to remain problematic, but not to the degree that the issues should preclude moving ahead with national commitments under the FCCC.

The IPCC has also encouraged countries to submit their inventories by the end of 1993. As of October 1992, 31 countries had submitted preliminary, often partial, national inventories. This progress will considerably enhance a prompt implementation of the FCCC.

Status of Estimation Methods for CO_2, CH_4 and N_2O

A vast amount of information is required to prepare national or global greenhouse gas inventories. The summary below focuses on the uncertainties in country-by-country estimates of emissions and removals of GHG, by source or sink category and by gas. It is based on reviews by the OECD, IPCC and US EPA. The cited ranges of uncertainty are necessarily generalizations, based on existing estimates. For any given country, the accuracy of estimates may be much better or worse than estimated here.

CO_2: Energy combustion accounts for some 60 to 85 percent of human-related emissions. Available methods are sufficient to estimate these within approximately 5 percent accuracy (IPCC 1992b), at least for countries with reliable energy data. The emission coefficients for certain energy products need to be reviewed and revised, according to IPCC's analysis, particularly for coke, natural gas liquids, biomass, and blast furnace gas. More important uncertainties are: whether all sources of CO_2 are included, to whom to attribute certain 'off-shore' emissions (e.g. from bunker fuels), whether emission coefficients are reflective of country-specific fuel supplies, and whether non-energy uses of fossil fuels are properly accounted.

Land conversion may account for very roughly 15 to 40 percent of human-related emissions. The range of uncertainty is probably no better than $\pm$ 50. The related removal of CO_2 by human land management is poorly estimated. So the net contribution of lands as a human-related source or sink may not be even directionally correct in some current, national inventories. Continual improvements in available methods and data should be expected over the next two or more years. An outstanding issue is an agreed definition of what will be considered 'anthropogenic' versus 'natural,' given the high degree of human interventions in natural systems in virtually all countries. High resolution remote sensing, combined with field studies for "ground truthing" and measurements of carbon per hectare, are critical to improving estimates of greenhouse gas fluxes from the forests and land cultivation. (See a related section in Part IV.) A substantially improved method is published in the mid-1993 report of the IPCC on estimation of GHG sources and sinks, but one should expect that this will undergo further rounds of refinement over the next few years.

Cement is a small contributor to anthropogenic CO_2, at one to two percent of the global total. Current methods are sufficient for estimating this source, given its small size.

Other sources and sinks of CO_2 have been identified, but their magnitude is poorly known and is probably not very significant globally. However, for individual countries, some of these sources may be important. These sources include: releases from oil and gas wells, coal mines, aluminum manufacturing, etc. Accounting of sequestration of carbon in long-lived forest products, such as wood housing, or landfills is rarely included in inventories.

All the percentages given in the following are assuming a mid-range total of 360 Tg of human-related methane emissions annually. The ranges in parentheses reflect the uncertainty of the source estimate, if 360 Tg were the anthropogenic total. Estimates are taken from IPCC (1992a).

CH_4: Enteric fermentation from livestock production may contribute 22 percent (19-26 percent) of human-related methane emissions. The methods of estimation used here are probably the best accepted among methane sources, and national statistics are fairly reliable. Other parameters, such as animal weights, age distributions, etc. need improvement.

Animal wastes may be about 7 percent (6-33 percent) of the anthropogenic contribution to methane emissions. Recent data will allow inclusion of a temperature parameter for wet waste management. Some countries have reported observational data up to ten times lower than the Casada and Saffley approach to estimation would produce. Additional measurements are needed to further improve the existing emission coefficients.

Coal bed releases of methane may be roughly 7 to 13 percent of anthropogenic methane emissions. Appropriate methods for estimating this source are understood, but more measured data are needed. However, not all estimates include the full scope of processes and post-mining activities. Also, data on coal mines for individual countries are often insufficient to apply the most appropriate rates of emission. Rates of methane capture for energy utilization may become an increasingly important parameter to include.

Oil and gas systems are among the most uncertain of sources, roughly 8 to 21 percent using "bottom-up" estimation methods. While production and distribution companies may have better activity and equipment data, national governments usually do not have sufficient information to estimate methane leaks from these systems to better than $\pm$ 50 percent accuracy. Systems vary too much from country-to-country to extrapolate accurately from existing national estimates.

Rice production is another highly uncertain source, although measurements are improving understanding of likely ranges of rates of emissions. This category may contribute roughly 17 percent (6–33 percent) of global anthropogenic methane emissions. However, uncertainties about the processes generating the emissions are still poorly understood, hindering precise estimation of fluxes.

Soils are a currently neglected category of removal or emission of methane in GHG inventories, although natural and human-managed soils together may remove some 30 Tg of methane per year from the atmosphere. No method has been proposed as yet to estimate the net anthropogenic flux in national inventories.

Landfill emissions of methane can be calculated with the existing methods, at about 8 (6 to 18) percent, but the range of uncertainty is large, due to lack of sufficient data about the magnitude and composition of wastes, the portion of potential methane production removed by soils prior to emission, and other environmental and management variables. A campaign of calibrated measurements would be necessary to improve significantly the methods. Also, the assumed time path of emissions relative to the time at which wastes are deposited varies substantially in different methods being used, complicating development of an agreed method for inventories.

Other human wastes may be a significant but poorly quantified source of methane emissions, at about 7 percent annually of the total. While emission factors are available in Japan and the U.S., this small source is usually not included in GHG inventories at present.

Biomass burning is highly uncertain at roughly 11 (6 to 20) percent of the human-related total methane emissions. The uncertainty is due to lack of sufficient knowledge about amounts burned and of measured data to derive reliable emission coefficients, rather than to uncertainty about an appropriate algorithm for estimation.

N_2O: Soils, including fertilizer use, may be the largest source or sink of N_2O affected by human activities. The processes determining the fluxes of this gas from soils are becoming better understood and models are being developed to estimate the net fluxes under a variety of soil, climate and management systems. However, too few measurements are available to validate the results of the new methods. Within this category, only fertilizer-related emissions are usually included in national inventories. Improving methods for this category should be a high priority.

Adipic and nitric acid production, associated with nylon and fertilizer manufacture, respectively, are newly quantified sources of human-related N_2O emissions. They are small in relation to the global total. Moreover, these are likely to diminish as sources as the production process changes to avoid emissions are relatively simple and low-cost. Methods are not widely available or validated for estimating the sources' strengths.

Fossil fuel combustion is the second largest among anthropogenic sources of N_2O. A measurement anomaly discovered in the mid-1980s caused earlier emission coefficients to be unreliable and far too high. Only a few new studies have been completed to determine accurately how emission rates vary with type of combustion (e.g. for fluidized bed combustion) or with other parameters (e.g. use of three-way catalysts for vehicle exhausts).

Biomass burning, particularly in tropical latitudes, is another important human-induced source of nitrous oxide emissions. New measurement studies have been carried out in recent years, improving the emission co-efficients associated with burning in different conditions. Also, the latitudinal gradients of N_2O concentrations in the atmosphere help to identify this source's contribution to the global total. (See section IV on use of ambient monitoring to improve source and sink estimates.)

Other Methodological Issues

Besides the issues identified above, there are a number of general priorities for improvements to the existing methods to prepare GHG inventories:

- Issues concerning the accounting for the release (or uptake) over time by certain activities in a given year remain unresolved, for example, concerning emissions from landfills or land clearing.
- Most of the emission factors available for calculating emissions come from measurements in the industrialized countries, especially the United States. Many of these emission factors are inappropriate for other countries with very different technologies, climates, and other relevant parameters. Therefore, a high priority will be on increasing measurements of emissions in a wider variety of countries and circumstances, in order to develop more representative

emission coefficients. (See "International cooperation to harmonize measurements" in Part IV.)

- Many countries will have to improve the coverage or quality of their statistics of economic and human activities, upon which the inventory methods rely. For example, more important than variations in the method for CO_2 from energy combustion, will be the quality and coverage of the underlying energy data on which the estimates are based. [Caution should be taken to watch that monitoring of emissions through economic data does not have the effect of pushing activities out of measured markets, for example from commercial energy into self-harvested biomass, whether explicitly for the purpose of hiding emissions, or as spontaneous avoidance of incentives (e.g. emission fees) to reduce emissions.]
- Interdependencies of emission rates of various GHG from certain sources are often not adequately analyzed. That is, changes in management or other parameters to reduce one gas may result in an increase or decrease of one or more other greenhouse gases. This suggests that improving the estimation methods, especially under emission control regimes, will require more focus on simultaneous measurement and modeling of multiple gases, rather than a single gas focus, as is now most often the case.

Training and Assistance

One point is clear from the IPCC work on GHG inventories: different countries have vastly varying capabilities and information with which to estimate their sources and sinks of greenhouse gases. The industrialized countries will generally be able to prepare detailed and fairly sophisticated inventories, while many of the developing countries will initially use simple methods relying on very limited data and assumptions generalized from other countries.

The IPCC has carried out a number of training workshops for participants from developing countries, and is developing simple software to help countries to prepare their inventories. In addition, UNEP, UNDP, the Asian Development Bank, and the World Bank are assisting developing countries to prepare their national inventories in accordance with the FCCC. The UN International Training Agency is also making plans to support the effort.

There will exist a great opportunity to use scientific collaboration on emission measurement and modeling to develop expertise in developing countries. This will also serve to improve existing GHG estimation methods. An interesting example of combined training, scientific research, and data development for policy is a project undertaken by the Mexican government to assess land use changes in the southeast of

Mexico. This project relies on a consortium of government officials, academic researchers and NGOs, and technical and financial support from the US government, to gather and analyze information which will be needed for that country's GHG inventory.

Reporting Standards and Review

The differences in national capabilities, resources and data availability suggest that "comparable" for the purposes of Article 12 will not mean "identical" among all countries. This, in turn, indicates that "transparency" will be an important means to assure the reliability and comparability of national estimates.

IPCC may recommend that an operational definition of "transparency" be that every emission inventory be accompanied by textual description of the methods used and references sufficient for a third party to reconstruct the emission and removal estimates from national activity data. The key assumptions that would have to be provided include emission factors, activity data and other variables used in the calculation algorithm (e.g., percent of methane in biogas from landfilled waste). A report in mid-1993 from an IPCC experts' meeting will identify such parameters. To minimize the task of documentation and to increase the consistency of national inventories, especially definitions, countries could use internationally available methods and data sets, making reference to them and identifying where they were used.

The OECD Secretariat, after comparing countries' initial and preliminary submissions in 1991, concluded that large differences are sometimes apparent between national economic data and international statistics. The international statistics are reported by countries to the pertinent international organizations, which adjusts them to make definitions, units, etc. consistent among all countries. For instance, a US-Mexican team working together on GHG inventories noted recently that using, variously, Mexican fuel survey data, IEA energy data, and UN energy data for Mexico, results in CO_2 estimates which differ by several percentage points (omitting biomass fuels, which the UN does not report). It was not clear why these statistics should vary so much nor which set is most accurate.

A common reporting format is being explored through the IPCC with a large number of industrialized and developing countries. A key element is to promote identical definitions of source and sink categories among countries. Another is to facilitate the compilation of national submissions for the Parties to the Convention, with disaggregation on a level of detail pertinent to policy making. For example, if a national government were to report measures to reduce GHG from, say, residential heating units, an estimate of emission specifically from that category would be useful to judge the likely effects from that mitigation measure.

The 'verification' by the OECD Secretariat of preliminary inventories has led to queries, and sometimes technical assistance, which have improved national estimates in a number of cases. Thus, some central cross-checks of data are probably efficient and useful in implementing Articles 4.1(a) and 12.1 of the FCCC and would promote the consistency of compiled national inventories. The OECD Secretariat has noted that the resources required to perform a simple check of national reports would be reduced to the degree that countries follow recommended guidelines on methods and reporting formats and provide adequate documentation. When other procedures are used, much of the Secretariat's time is spent making conversions of national data into similar definitions and units, increasing both costs and the potential for errors in the comparison.

An experimental "transparency" process, led by the OECD and IPCC, has paired off countries to review each other's preliminary GHG inventory. This has provided useful comparison and feedback on the completeness, intelligibility and accuracy of the initial submissions. The Australian-U.S. report (unpublished) to the IPCC workshop on October 1, 1992 suggested that:

- A complete set of background activity data, or reference to an internationally available, published source of data, is necessary to reproduce and validate a country's inventory.
- Reproducibility of inventories will become more important to evaluate future changes in reported emissions and removals, and to understand the effectiveness of national policies and measures in effecting those changes.
- Countries should be able to explain differences between the data sets they use for their inventories, and those published by the UN (or the OECD/IEA), if they differ significantly.
- Emission factors should be reported at the same level of detail as the incidence of national policies and measures, in order to provide useful feedback on the effectiveness of national policies and measures.
- Scientific studies from which emission or removal co-efficients were derived should be published and internationally available, and referenced in the inventory report, to assist comparison with those used by other countries.
- If countries use definitions other than those recommended by the IPCC, they should be clearly stated, and an effort made to indicate the correspondence between the two definitions.

- For activities and gases where considerable inter-annual variability exists, a multi-year averaging method might provide more reliable results than an estimate for a single year.
- Reporting of ranges of uncertainty is useful, as well as some explanation of how the range was derived (different methods, data sets, emission coefficients, etc.).

Generally, it was productive both to review another country's inventory as well as to be reviewed. It also required relatively few resources from a central secretariat. Such a decentralized, expert approach could rely heavily on scientific collaboration and peer review. This, rather than a strong central bureaucracy, may be a useful model for the Parties to the Convention to consider.

National Policies and Measures to Mitigate Greenhouse Gases

All Parties to the FCCC commit themselves to "formulate, implement, publish and regularly update national, and where appropriate, regional programmes containing measures to mitigate climate change by addressing anthropogenic emissions by sources and removals by sinks of all greenhouse gases not covered by the Montreal Protocol, and measures to facilitate adequate adaptation to climate change." [Article 4.1(b)] Developed country Parties and those listed in Annex I commit to a somewhat stronger formulation which recognizes the contribution that reducing net emissions from current levels would make to the goal of the Convention [Article 4.2(a)]. Beginning within six months after entry into force of the Convention, these Parties will communicate in detail their policies and measures, including projections of their expected effects. The Conference of the Parties will promote and guide development of comparable methodologies, including those for evaluating the effectiveness of measures to limit the emissions and enhance the removals of these gases [Article 7.2(d)]. Furthermore, the Conference of the Parties will assess the overall effects of the measures and the degree of progress towards the objective of the Convention [Article 7.2(e)].

The process could be divided into two parts: (1) reviewing individual countries' policies and measures, allowing a diversity of methods, as long as they follow reasonable or best practices, and (2) a mutual, integrated assessment of the cumulative effects, looking at international effects, and allowing evaluation of overall adequacy. The latter activity would have to use one or more common frameworks for analysis, presumably quantitative, into which each country's policies and measures would have to be approximated and entered.

Forecasts of Future Emissions

Both developed and developing country Parties have committed themselves to provide projections of future emissions, or the information with which to do so (if feasible), in Articles 4.2(b) and 12.1(c), respectively.

The forecasting provisions of the FCCC are important for at least two distinct purposes. First, forecasting the cumulative effects will be the foundation for evaluating the adequacy of the commitments by the Parties. Second, determination of the "agreed full incremental costs" of implementing measures, for which developed country Parties have committed to provide financial resources to developing country Parties, will rely in part on projections of cases with and without national actions. Each of these purposes will be discussed below.

Assessing the Adequacy of Commitments

Reporting and forecasting the effects of policies and measures is critical to the ability of the Conference of the Parties to evaluate progress toward the long-term goal of stabilizing concentrations of greenhouse gases in the atmosphere. At the time of submissions, there will be a strong need to evaluate the trajectories[3] of greenhouse gas emission and removal to indicate any discrepancy between the nationally reported trajectories and the global stabilization objective.

In future years, countries' GHG inventories can be compared against their previously projected emission levels. This could provide an 'early warning system' as to whether the national policies and measures, as carried out, are achieving their anticipated effects. It would assist improvements in both the forecasting techniques, and possibly adjustments in the policies and measures themselves. This review could be cross-checked with ambient measurements by the scientific community. (See also the section on ambient monitoring in Part IV.)

Strictly to assess the adequacy of the Convention, developing a baseline without climate policies and measures, is neither called for, nor absolutely necessary. It is the projections with policies and measures that matter and the comparison against commitments. Also, economic evaluation of national policies and measures for developed countries is not required under the FCCC. Economic evaluation would be important, however, if the FCCC were to strive toward equalizing the marginal levels of (economic) effort among countries to mitigate GHG, but that is not now the case. Such economic evaluation may nonetheless be a useful element of collaboration.

Determining "Agreed Full Incremental Costs" of Measures

In the case of developing countries seeking financial assistance, both baseline projections and economic evaluation of policies and measures become implicit requirements of the FCCC. Although the commitments to provide forecasts from developing country Parties are weak, and do not explicitly require economic evaluations, such information will undoubtedly be needed in order to implement the financial assistance clauses of the Convention.

"Incremental" costs are those costs which would be incurred by a proposed program above (or below) a baseline of what would be incurred if the mitigation program were not undertaken. Wilson (1992) points out the importance yet impossibility of determining an objective baseline from which to calculate "agreed full incremental costs." He cites an internal World Bank paper by King and Munasinghe which identifies several analytical problems which can only be solved subjectively. Wilson quotes from the latter analysis, noting that "any incremental cost calculation done to satisfy the needs of an international compensation process 'will always remain hypothetical...because the baseline investment program remains counterfactual.' Thus, they assert, the strongest incentive for the developing country would be to 'make a "good" baseline plausible rather than to try to control actual expenditures.'

Thus, although the FCCC does not contain explicit commitments for developing country Parties to provide baseline or projected emissions, this will nonetheless become an important and probably controversial component of implementation.

Information to Include in National Reports

The types of information probably necessary to report with national policies and measures will include:

- *Demographic Facts and Assumptions:* Birth rates, death rates, migration, age distributions, household sizes, and changes in community size distributions (i.e. urbanization);
- *Economic Structures and Trends:* Historical trends, reasons for expected breaks in trends, income per capita and distribution within the economy; activity levels by sector; structural changes within and among sectors; changes in lifestyles, consumer preferences; cultures and traditions; and expectations about resource availability and prices;
- *Economic and Social Policies:* Legal and institutional policies toward ownership, competition, price-setting, taxation and subsi-

dization, openness to international trade. This information would be most useful if provided by sector, including energy, agriculture, forests and public land management, and government services (e.g., provision of infrastructure, waste management, etc.).

- *Detailed Descriptions of Policies and Measures to Mitigate Greenhouse Gases and to Enhance Sinks:* General approach or philosophy; specification of types of instruments employed (e.g., taxes, subsidies, regulations, training, direct investment, communication campaigns, volunteerism, etc.); to what the instruments apply, and their stringencies, related to the previous policies (i.e. changes in tax rates relative to the base); and the change expected to be induced by each program.
- *Quantification of Effects:* Either for each action or cumulatively, recognizing that a package of actions may have greater or lesser effects than the sum of each individually.
- *Methods Used to Quantify Effects:* The type(s) of analytical framework or instrument used, with an explanation of its strengths and weaknesses; an assessment of the major uncertainties and the range of uncertainty associated with final estimates.

A point of note is a clause in Article 12.9 which allows Parties to designate information as confidential, according to criteria established by the Conference of the Parties. Caution should be taken in defining strictly what could make information "confidential," as too loose a definition could substantially hinder both scientific and analytical exchange among the Parties and researchers. While an argument exists in favor of allowing confidentiality to maximize the information provided by Parties, the U.S. GAO (1992) notes that the secretariats to environmental agreements cited do not have the resources to review and verify in detail the reports of participating countries. If submitted information is not made available to Parties, it may constrain implementation by decentralized mechanisms, such as expert working groups, rather than by a centralized international bureaucracy.

Comparable, Not Identical, Methods

Countries which will be Parties to the Convention will have greatly varying economic and social structures. These different structures will be more amenable to analysis with some types of methods than with others (witness the implausibility of using market equilibrium models to emulate the transitions now of the economies of Eastern Europe and the former Soviet Union). Moreover, each type of action – whether projects, programs or broad policies – may be more adequately evaluated with certain analytical approaches than others.

For example, an investment in, say, methane capture and utilization on a landfill, would be better analyzed with cost-benefit analysis than with general equilibrium modeling of the entire economy. In many countries or situations, the method of analysis will likely be determined by the availability of expertise and data. In any case, there are no perfect analytical approaches or tools: each has its usefulness in answering certain kinds of questions. A UNEP project, Preparation of a Methodology to Undertake National Greenhouse Gases Abatement Costing Studies, concluded recently that "capturing all the elements ... is far beyond the capability of any computer model at present; in practice, different models are designed for different purposes, they have different strengths and weaknesses, and yield different information. Modelling of some aspects, particularly technological development and the circumstances of developing countries, is still especially weak." (UNEP 1992) For all the reasons given above, the possibility or the desirability is doubtful of attempting to recommend a specific method to evaluate national policies and measures.

However, a useful objective for the Conference of the Parties may be to establish guidelines as to "best practices" for the range of analytical tools available and likely to be used to evaluate national policies and measures. Even Wilson's (1992) contention that "even the most careful and rigorous analysis is not an objective finding" is not incompatible with an attempt by the COP to define what consititutes "careful and rigorous" analysis. Standards for scope and types of analysis could help to improve and make comparable the information reported by Parties.

Despite the limitations inherent in any analytical approach to measuring the effectiveness of national policies and measures, "best practices" could be outlined for a variety of analytical frameworks, including:

- cost analysis of specific projects or programs;
- simulation analysis (sometimes called "engineering analysis" or "bottom-up") of a set of programs and measures;
- incremental cost curves, usually based on micro-economic evaluation of projects, programs or policies;
- optimization modeling of a project or sector;
- input/output analysis of an entire economy;
- dynamic, equilibrium or partial equilibrium analysis of broad policies and programs; and
- international equilibrium analysis of the policies and programs of a set of countries or regions.

The UNEP project mentioned above, through Risø Laboratory in Denmark, has undertaken a useful process to review some of these specific methods, and to test and recommend a common process for national cost assessments. Their Phase I report concludes that "short-run models are in general still far from adequate to give reliable assessment of the full range of options" to mitigate greenhouse gases.... Estimates derived from long-run models are highly uncertain because of profound uncertainties concerning technical change and the likely evolution of energy systems and fossil fuel markets even in the absence of abatement policies. Such assessments ignore both transitional costs and the potential for 'zero cost' savings." (UNEP 1992) Despite the uncertainties and limitations inherent in current analytical methods, they still can yield very valuable information for decision-makers about the types and magnitudes of effects to expect from different policy or investment options.

Common Assumptions and Integration of National Reports

There will be a need to establish some principles or common assumptions for national analyses, if the cumulative effects are to be evaluated, even if countries choose to use their own alternatives for their official reports. The key parameters will include sectoral or activity definitions; common reporting units; assumptions concerning global population trends, including migration; global economic growth rates and regional distributions; supplies of energy and agricultural resources and international price trajectories; and perhaps some assumptions regarding the availability and prices of key technologies over time.

After countries report their policies and measures, and an international assessment is developed from them, it may be necessary to iterate the analysis back to national policies with changes in assumptions, such as those regarding international prices.

International evaluation of the cumulative effects of national policies and measures is important. It will help to assess possible "leakages", aggregation of markets and resulting price effects, and changes in trade and economic growth globally, for countries and for blocks of countries.

Development of methods for effective review of national policies and measures will probably take some time. One should expect several trials and errors before an effective and acceptable procedure could be recommended to or by the Parties.

A core group of countries could begin to test some alternative processes prior to the first meeting of the COP. It might include particularly those countries with policies and measures already reported and willing to submit themselves to review. As workable procedures are identified, and the set of countries with reported policies and measures expands, the participating group of countries could broaden, eventually to include all Parties. Reports on progress could be made regularly to the

Subsidiary Body for Implementation and to the Conference of the Parties on interim findings.

The value of the review process for building of expertise and institutional capacities in all countries should not be overlooked. As is the case in the greenhouse gas inventories, the opportunities for training and exchange of information should be built into a review process. As the report on Phase I of the UNEP project on abatement costs concludes, "Comparison of different national costing studies could enhance the learning process and point to new and lower cost abatement options." (UNEP 1992)

Variations in Cases of Joint Implementation

The FCCC allows countries to achieve individually or jointly their commitments under Article 4.2(a). The COP in its first session will have to develop criteria for permissible "joint implementation." In the simplest of cases, joint implementation could be simply that two or more countries (for example the EC) together meet their cumulative commitments under the FCCC. This presumes a closed system among the countries involved and that the national inventories of the participating countries would cumulatively be compared against the sum of their national commitments.

An alternative form of joint implementation is likely to be sought under the FCCC. It would entail agreements in which one country (or an entity within one country) pays for some action or program in another country and, in exchange, receives credit for some portion of the resulting emission reduction or sink enhancement.

Joint implementation in the latter form is likely to require additional information beyond that required by individual emission limitation commitments. More specifically, while an individual national inventory of emissions and uptake could be documented with reasonable transparency using aggregate statistics and generalized emission factors, approval of credits for joint implementation will require site-specific estimations as well as analysis demonstrating that emissions are not simply being shifted to another site. It may also require a demonstration that those greenhouse gas reductions or sink enhancements would not have been achieved without the joint implementation project.

For actions involving sources of emissions, measured emissions and activity levels would be necessary to establish the emissions in the base year. The task would be more complex for a joint implementation project involving a sink, such as an afforestation project. Since the uptake cannot generally be measured directly, other types of data would be needed to determine the level of uptake incrementally achieved by the project. This might require, for example, remote sensing images of the location (over time), sampling of the vegetation and soils, determination of the history of the site and the maturity of the ecosystem in place, specification of the growth rates of the species of vegetation before and with the project, collection of local temperature, rainfall and other data, and modeling of the expected changes in the site as a result of the project.

In addition, it would be necessary to determine a baseline for the site and region or country, showing how the level of emissions or uptake would increase or decrease over time in the absence of the joint implementation project. Establishing this baseline might well be the most controversial and difficult step of an approval. However, it would be necessary to ensure that the project is not counting changes that would have occurred without the project, and to show that the emissions or degradation of sinks are not being shifted elsewhere.

It may be especially difficult in many cases to avoid double-counting the effects of a joint implementation project and effects of other policies and measures carried out within the host country. As an hypothetical example, how would one assess the credits allowable to a company which pays to install compact fluorescent light bulbs in a foreign installation when, simultaneously, the local electric utility initiates a demand-side management program aimed at industrial consumption? Determining how to allocate credit for such reductions may be difficult, and raises the issue of who will have authority to approve a joint implementation project.

Several mechanisms could help to resolve these issues for the COP. First, a discount could be required of emissions credits gained through joint implementation, to account for an increased risk (real or perceived) that the emission reduction or sink enhancement will not be maintained. This might vary with the type of project, the past performance of participants, or other factors. Second, the body with authority to approve credits for joint implementation could require a bond or other guarantee of the right and capability of the entity supplying the credits to sustain the emission reductions or sink enhancement over time. Third, rigorous measurement of emissions following implementation of the project may be required, to compare against the expected change. More frequent and thorough field monitoring might be required in cases of joint implementation than that needed for Parties which meet their commitments individually.

Fourth, placing liability on the country providing the greenhouse gas credit (i.e. the party which receives payment in return for reducing emissions or enhancing uptake) would instill a greater incentive to avoid double-counting, to ensure that the emissions or the sink degradation are not shifted to another location within the boundaries of the provider of the greenhouse gas credit, and avoids problems of sovereignty in enforcing the effectiveness of the project.

One mechanism which would avoid a centralized bureaucracy, with a large burden of review and reporting to the COP, would be to decentralize the process. In a decentralized system, the COP would establish rules under the FCCC for evaluating and certifying joint implementation projects. The country providing the credits (itself or through some entity within its borders) would assume liability for the greenhouse gas reduction (by effectively making its commitments under 4.2(a) more stringent), regardless of whether the project were contracted by a public or private entity. The country assuming liability would have authority to certify the projects and responsibility to ensure that it is carried out. That country's baseline commitment under the FCCC would be adjusted accordingly. The primary information the Conference would need, then, would remain the communication of national policies and measures, along with central accounting of which countries have bought or sold credits. Compliance would be measured in future years by comparing each country's greenhouse gas inventory against its adjusted commitment. This system would be meaningful only among countries which have adopted specific greenhouse gas limitation commitments (currently Annex 1 only).

Scientific Cooperation and Exchange

Scientific cooperation and exchange of information under the FCCC can be designed and used to assist the other provisions of the convention, including those on greenhouse gas inventories and national policies and measures. These scientific activities can improve the data available for national inventories and policy-making; increase the comparability of information reported by individual Parties; help to verify or improve information communicated by Parties; and provide essential training, especially for technical personnel from developing countries.

Improve Economic and Social Statistics

Improvement of national and international statistics on important human activities, for example on energy balances, agricultural production, or changes in land uses, would be one of the single most useful steps to improving emissions and uptake information and the ability to verify national reports. It would in most cases constitute a high priority, "no regrets" measure for countries, since such data would help to monitor and enhance economic development. The importance of the veracity of economic data for other uses would be a disincentive to manipulate them to show compliance with commitments in the climate convention.

TABLE 3.1 Comparison of Satellite Imagery

Satellite	Frequency of Overpass	Resolution (pixel)	Coverage per Scene	Cost per Scene	Cost $/10^6 ha$
AVHRR	Twice daily	1.7 km^2	90 X 10^{6}ha	$ 130	0.7
Landsat MSS >2 scenes/yr.	16 days	0.6 ha	3.4	$ 200	60.0
Landsat MSS <2 scenes/yr.	16 days	0.6 ha	3.4	$1000	294.0
Landsat TM	16 days	0.09 ha	3.4	$4400-5500	1294 to 1617
SPOT	26 days	0.04 ha	0.36	$2450	6805

Source: Houghton 1992
Reprinted with permission.

Increase Use of Remote Sensing and Field Studies

Remote sensing, particularly the use of satellite imagery (see Table 3.1), for detecting and monitoring changes in ground-level activity, such as forest clearing or plantation, as well as the distributions of concentrations of certain greenhouse gases, is at present more in the realm of research than policy analysis. Improving the coverage and consistent interpretation of remote sensing is probably among the highest priorities for scientific collaboration under the climate convention in order to improve information for policy-making.

Various researchers have outlined the research needed to substantially improve GHG fluxes associated with land conversion (Houghton 1992 and Skole et al. 1992). Their proposals are consistent and include the following components:

- enhancement of the coverage of remote sensing data, by adding or upgrading satellite receiving stations;
- calibration of measurement and interpretive methods internationally;
- greater use of Geographic Information Systems, especially internationally, to accumulate and 'overlay' site-specific data;
- "wall-to-wall" measurement of baseline land uses, and change detection, using high resolution but low frequency (3 to 5 years) satellite imagery (e.g. Landsat MSS);
- monitoring of fires with high frequency satellite imagery (i.e. AVHRR, close to daily during dry seasons)
- annual, very high resolution sampling of key regions (using Landsat TM or SPOT)
- calibrated, ground-level measurement of carbon content of vegeta-

tion and soils, of land uses and management, and causes of changes detected by remote sensing;

- increased measurement of emissions related to land uses and conversion to improve emission coefficients;
- increased statistics on the "fates" of forest products, such as paper, timber, fuelwood, and on the timing of their subsequent emission of GHG or sequestration of carbon;
- improved models to calculate GHG fluxes over time as a function of natural processes and human interventions.

Use Ambient Monitoring to Cross-Check Inventories and Projections

There is a variety of reasons that discrepancies may appear between the levels of net emissions reported and projected by countries over time, and the actual accumulation of greenhouse gases in the atmosphere. In principle, concentrations should correspond with the national estimates of sources and sinks (plus natural sources and sinks). Reasons for discrepancies may include: missing sources or sinks; miscalculation of the strengths of sources and sinks, due to imperfect knowledge or data; or misrepresentation. Another reason may be imperfect understanding or ability to quantify (model) the associated atmospheric processes of reactions, removals, etc.

Monitoring of ambient concentrations of greenhouse gases, and their spatial distributions, along with the isotopic signatures of some of the gases, could help to cross-check the accuracy of "bottom-up" emissions calculations and projections, at least at the regional level. For example, information about sources and sinks of methane, and their regional distributions, are intrinsic in the north-south gradients, seasonal and geographic variations of methane concentrations, especially when the variations in isotopic composition are evaluated. (Matthews 1992).

Ambient monitoring on a regional scale tends to integrate emissions from different sources. Fung et al. (1993) use the isotopic concentrations of CH_4 to detect the fractions of fossil and biogenic carbon in atmospheric methane. This has proved very useful in placing bounds on the contributions of different source types in world regions to the global annual total emissions. This work played a significant role in determining the global methane budget for the 1992 IPCC Scientific Assessment. If developed further, it could assist implementation of the Convention.

The resolution of the monitoring network would partially determine the size of the region for which total, net emissions could be interpreted with this "top-down" approach. The current US NOAA/Climate Monitoring and Diagnostics Laboratory supports a cooperative network of long-term monitoring sites which were chosen to represent background concentrations relatively unaffected by emissions from human activities. Additional sites would have to be identified in order to reflect changes in emissions or removal, rather than background concentrations.

Measurements of long transepts by aircraft or ships can also be used to supplement a fixed monitoring network. They can place constraints on the strengths of sources upwind. Significant improvement in the existing monitoring networks, dispersion modeling and interpretation would be necessary and difficult, but possible. Matthews (1992), for example, notes that, to narrow the uncertainties in the methane budget, it would be necessary to:

- improve the accuracy and coverage of 14C and 13C measurements in the atmosphere and from sources;
- expand three-dimensional atmospheric trace gas measurements to SE Asia, tropical South America, Africa and boreal habitats;
- further develop photochemical-transport models to synthesize and evaluate suites of source and sink distributions.

A critical issue to be answered is whether the current or expected future accuracy of the measurements, combined with dispersion modeling, will be sufficient to provide useful monitoring and assessment of the changes in net emissions sought under the FCCC.

Enhance International Cooperation to Harmonize Measurements

Calibration of measurements is critical to obtaining results which are consistent internationally and over time. This is as important for such tasks as interpretation of remote sensing or estimation of volume of carbon per hectare, as for directly measuring rates of emission or removal.

Periodic inter-calibration of measurement techniques will be desirable among the various researchers internationally engaged in monitoring emissions from sources and concentrations in ambient air. Analogous procedures are being established for interpretation of remote sensing under the IGBP, and could be developed for in situ measurement of carbon in standing vegetation.

"Round-Robin" measurement experiments among laboratories in the European Communities, and in the United States, have been a useful component for quality assurance in the respective air pollution control programs. These involve measurements of identical samples by different laboratories and comparisons of results.

Develop Models to Target Monitoring Needs

An improved link between scientific disciplines and policy development could make the most effective use of resources to implement the

FCCC. For example, development of new spatial-economic models, linked to Geographic Information Systems, could help to target those areas most susceptible to changes which increase or remove greenhouse gases in the atmosphere. This information, in turn, would be used to stratify samples, to target remote sensing, or to establish field studies or monitoring of activities on the ground. This could help to reduce the potentially high cost of scientific activities, and assist efforts to comply with requirements of the FCCC.

Spatial economic models will also be important in evaluating the cumulative effectiveness of national policies and measures. Countries will want to be assured that their programs and projects designed to reduce emissions or to enhance removal of emissions have the desired net effect and do not just shift the activity to another location.

Develop a Strategy for Key Research to Support the FCCC

One near-term activity under the FCCC could be the development of a longer-term sampling strategy for ambient and field measurements of various types needed to verify changes in emissions and atmospheric concentrations over time. This would involve ambient measurements, selected source measurements and other field studies, remote sensing, improvement of ambient modeling and synthesis techniques, etc. Priorities for improving the methods and inter-calibration internationally and international training would be specified. The needs for temporal and spatial coverage would be determined, as well as how these needs could be met. The priorities and potential allocations of responsibilities could be proposed, including the important roles that independent research institutes and non-governmental organizations could play.

Conclusions

Effective implementation of the Framework Convention on Climate Change will require the development, reporting and analysis of a wide variety of information. At the heart of the information needs is the requirement that all Parties to the Convention must compile and report national inventories of greenhouse gas sources and sinks. While a large range of uncertainty will surround many of the estimates for several years, the quality of methods and information available is sufficient to proceed with other aspects of implementation of the FCCC that will rely on the inventories.

The fundamental commitment of all Parties is to develop (and report) national policies and to implement measures to mitigate greenhouse gas emissions and to enhance sinks. National plans will need to reflect the general commitments of Parties in Article 4.1, such as to promote practices and processes that control greenhouse gas emissions and to promote sustainable management of sinks and reservoirs. They will also need to reflect the specific commitments in Article 4.2 of Annex 1 countries. It would be impracticable for the COP to prescribe the methods countries must use to report the effectiveness of their policies and measures. Nonetheless, identification of best practices for different analytical methods could facilitate comparability of countries' evaluations. It also would aid assessment of the cumulative effectiveness of national policies and measures toward the objective of stabilizing atmospheric concentrations of GHG.

The provision which allows Parties to meet their specific commitments jointly will require greater and more specific information. "Joint implementation" will require site-specific information, projections of activities and greenhouse gas fluxes with and without the project, and assessment of how the project might affect fluxes in the surrounding country or region. This is necessary to avoid counting credits for what would have occurred anyway, to avoid double-counting, and to avoid "leakages" of emissions by shifting activities to other locations. Assigning liability for the abatement project to the country which hosts the project (i.e. which "sells" greenhouse gas credits) could minimize the necessary international review of projects.

Finally, the scientific collaboration provisions of the FCCC could be used to help carry out countries' mitigation commitments. Specifically, scientific collaboration can help to verify on a broad scale the greenhouse gas inventories reported by countries. Such a comparison would utilize the inventories, and ambient and source measurements in association with dispersion modeling. Remote sensing can also help to track greenhouse gas fluxes over time, and changes due to national policies and measures. It could also help to target the most vulnerable land areas where policy attention may be most needed.

In conclusion, an interdisciplinary approach to meeting the information requirements of the Framework Convention on Climate Change will be required to carry out countries' commitments effectively. It demands collaboration among scientists, technical experts and policymakers. In addition, careful design of criteria and guidelines for reporting and review of greenhouse gas inventories, national policies and measures, and joint implementation, combined with development of "best practices" for analytical methods, could also enhance the effectiveness of the Framework Convention.

Notes

1. This chapter only addresses issues relating to mitigating greenhouse gases, not those relating to the impacts of climate change and adaptation.

2. Only for those gases not covered by the Montreal Protocol of the Vienna Convention to Protect the Stratospheric Ozone.

3. This discussion does not preclude the possibility that Parties will be allowed or encouraged to provide projections under a range of plausible assumptions, to reflect uncertainties both in the effectiveness of the policies and measures, and in evolution of the economic and social environment.

References

Fung, I., E. Dlugokencky and B. Braatz (draft). 1993. "Verification of Methane Emissions" in U.S. EPA, *Global Anthropogenic Emissions of Methane: A Report to Congress.*

Houghton, R.A. 1992. "A Blueprint for Monitoring the Emissions of Carbon Dioxide and Other Greenhouse Gases from Tropical Deforestation" presented January 20, Woods Hole, Mass.

IPCC 1992a. *Climate Change 1992: The Supplementary Report to the IPCC Scientific Assessment.* Cambridge: Cambridge University Press.

IPCC 1992b. Reported in the IPCC/OECD Workshop on National GHG Inventories Estimation and Reporting Transparency, Bracknell, UK, October 1.

Matthews, Elaine 1992. "Assessment of Methane Sources and their Uncertainties" draft to be presented to a meeting sponsored by the Russian Academy of Sciences/IUPAC/IIASA/ECN, 19-20 July, Moscow.

Skole, D.L., C.O. Justice and J.P. Malingreau 1992. "A Satellite Based Tropical Forest Monitoring and Information System: A Plan for Implementation" draft of 26 October.

United Nations Environment Program 1992. *UNEP Greenhouse Gas Abatement Costing Studies: Phase One Report* UNEP Collaborating Centre on Energy and Environment, Riso National Laboratory, Denmark, August.

United Nations General Assembly 1992. "Report of the Intergovernmental Negotiating Committee for a Framework Convention on Climate Change on the Work of the Second Part of its Fifth Session," held at New York from 30 April to 9 May: Addendum" A/AC.237/18 (Part II)/Add.1, New York, 15 May.

U.S. Government Accounting Office 1992. "International Agreements are Not Well Monitored," GAO/RCED-92-43 Washington, D.C.

U.S. Government Accounting Office 1992. "Strengthening the Implementation of Environmental Agreements," GAO/RCED-92-188 Washington, D.C.

Wilson, John D. 1992. "Financing Climate Mitigation in Developing Countries" JFK School of Government, Boston Mass., April 20

The views expressed here are solely those of the author and do not necessarily reflect those of the U.S. Environmental Protection Agency, nor the U. S. Government.

4

Climate Change Policies in Developing Countries and the Role of Multilateral Institutions

Eunice Ribeiro Durham

Introduction

The Rio meeting on Climate Change had a very ambitious objective which was, in principle, extremely difficult to attain: to reach an international agreement which would bind all sovereign nations of the world to a commitment to enforce, within their jurisdiction, polices directed to the control of climate change.

The general difficulties of such an agreement are of two orders. In the first place, it is certainly easier to establish agreements of this kind when the benefits which would derive from them are immediate, easily perceived and affect everyone in the same manner and with the same intensity. Clearly, none of these conditions were present in the attempt to draw a Climate Change Convention.

In the second place, international discussions which lead to agreements are easier when the partners are relatively equal; when they have the same responsibility for the phenomena which is to be avoided or for the objective which should be attained; when they have the same power and resources to enforce decisions, so none is able to unduly press the other partners to accept its own position; and when the sacrifices and benefits are equally shared. Again, the context of the Rio Climate Convention was quite different from this ideal.

Given the difficulties, it is somehow surprising that an agreement was finally reached. And, considering that the problems regarding opposing interests among unequal partners were not effaced, they certainly will influence the implementation of the convention and should be carefully taken into account.

International agreements unfold within the context of existing political, economic and military relations among nations. The Climate Convention was no exception and the issues involved were of much wider scope than climate change itself. It was in the long negotiations preceding the Rio meeting, rather than in the final discussion, that the political and economic conflicts of interest were clearly stated and perceived. And, although there were different cleavages so that on every issue allies and antagonists were not the same, undoubtedly the opposition between developed and developing (or underdeveloped, or plainly poor) countries dominated much of the debate.

This chapter discusses two different but associated questions. The first refers to the specific difficulties encountered by developing countries in the implementation of climate change policies. The other has to do with the role of multilateral institutions in the promotion of such polices in these countries.

In analyzing these questions, it is important to avoid interpreting the opposition between developed and underdeveloping countries as a battle of good against evil. An objective analysis must be based on a clear definition of the opposing interests so we can understand where lies the root of possible conflicts.

The developing countries' position rested on three main interrelated issues: the problem of responsibility for climate change, the relation between climate change polices and development and, most important of all, the question of financial aid for climate change control measures.

All these issues touch old and new grievances of developing countries vis-a-vis the rich and powerful nations. They have to do with the increasing difficulties faced by developing countries in diminishing the gap which separates them from First World prosperity. In many cases, indeed, the gap is increasing and, to a greater or lesser degree, poor countries attribute much of the blame for this situation to the economic world order which results from the dominance of developed countries. In view of the economic problems resulting from the trade balance and foreign debt, many countries feel that international financial aid is the only solution that would allow them to surmount their present situation of dire and, sometimes, increasing poverty. For many of these countries, developed nations' concern with climate change appears as the opening of new possibilities of access to international financial resources which could help them in their present difficulties.

The objective of this chapter is not to discuss the justice or injustice of the present world order. But it is necessary to acknowledge the existence of grievances because they certainly have conditioned much of the Convention's negotiations and will certainly influence future action. It is in the light of these grievances that we can understand the importance attached by developing countries to the three main issues pointed out above. It is also in the light of existing opposition that the importance of the role of multilateral institutions will become clear.

Development and the Responsibity for Climate Change

Surely, the threat to global environment presently associated with climate change is a result of the process of industrialization and technological development which characterized the history of modern Western societies, rather than a product of poverty or lack of development. It was in great part by increasing energy consumption, which is the main factor leading to the increase in CO_2 emissions, that modern industrialized nations reached their present position of wealth and power. And, as far as our knowledge goes, CO_2 is the most important factor in the greenhouse effect.

It is because they are developed that modern industrialized nations account for three-fourths of the emissions of gases responsible for the greenhouse effect, most of which result from burning fossil fuels. All together, developing and plainly poor (or non-developing) countries, which include 75 percent of the world's population, contribute only 25 percent to the total emission of the gases.

In the face of such evidence, it is easily understandable that developing countries should claim that climate change is mainly a responsibility of industrialized nations and that they should bear the greatest burden in the effort to reduce emissions.

The recognition by developed nations of their main responsibility for climate change was a crucial point in North-South opposition during the meeting in Rio, and it may be considered a victory of developing countries that article 4 of the Convention should state:

The developed country Parties commit themselves specifically as provided for in the following:
a) each of these Parties shall adopt national policies and take corresponding measures on the mitigation of climate change, by limiting its anthropogenic emissions of greenhouse gases and protecting and enhancing its greenhouse gas sinks and reservoirs. These policies and measures will demonstrate that developed countries are taking the lead in modifying longer-term trends in anthropogenic emissions consistent with the objective of the Convention, recognizing that the return by the end of the present decade to earlier levels of anthropogenic emissions carbon dioxide and other greenhouse gases not controlled by the Montreal Protocol would contribute to such modification...

Undoubtedly, the phrasing of the document sometimes sounds as if developed countries agreed to take the lead in restricting anthropogenic emissions conducive to climate change as a matter of unselfish regard for ecological preservation. The truth is that, given the volume of developed countries' emissions, unless they control it it would be impossible, no matter what developing countries do or do not do, to counterbalance the present level of CO_2 concentration, and even its increase.

In the present situation, and considering the impact they might have in decreasing the concentration of greenhouse gases in the atmosphere, climate policies in developing countries are clearly secondary and subordinated to analogous policies in developed countries. For this reason, the rationale for imposing strict rules on the emissions of industrialized countries – such as establishing a ceiling for their emissions to be reached in the year 2000 – is very compelling, and it was the policy of developing countries to press in this direction. It is certainly within their reach, both from a financial and a technical viewpoint, to achieve it. Some countries of the OECD have unilaterally adopted policies that will lead to such a stabilization of their emissions. In this regard, the United States' opposition to such measures not only diminishes the credibility of developed countries intentions but also threatens the very objective of attaining climate equilibrium. There are, then, different positions among developed countries, particularly between the European Community and the United States.

But there are also some important internal differences among developing countries. The main contribution of developing nations to the greenhouse effect is due to forest destruction and coal burning, which are heavily concentrated in a few countries. Only a few countries still have large forests so as to make forest burning a significant contributor to CO_2 emissions. Brazil is the most important of these, having the largest area of tropical forests in the world. Indonesia is another. China is also a great polluter, but for different reasons, related to methane liberation in inundated rice fields and coal burning. To these three, we must add India, owing to its sheer size and population. But at present, and for the majority of developing countries, large CO_2 emissions neither pose a problem nor constitute a possibility, and they may take a different stand regarding these questions.

It is therefore necessary to recognize that developing nations are not a homogeneous block. They have different problems, possibilities and perspectives for the utilization of natural resources and it is by no accident that the "polluters" are among the most developed of the underdeveloped countries. However, even considering the case of Brazil, China and Indonesia, it is still true that emission of CO_2 and pollutants by developing countries is much less significant than that of developed countries. It is then understandable and reasonable that the emphasis put by developed nations on the need to control forest and coal burning should appear to them as a way of avoiding the main issue: the need to control oil burning, an effort which should be mainly undertaken by developed nations. "Great Polluters", such as Brazil, China, India and Indonesia, are in a position to try and balance demands made upon them by developed countries with counter-proposals for their reduction of CO_2 emissions. And, since the policies in which they might engage themselves will have a potentially large impact on climate change, they are also in a strong position to request financial and technological support to help them implement such policies.

The fact that, among underdeveloped countries, the great pollutants are those which might benefit most from possible financial aid directed towards climate change control and which also have a strong bargaining position, gave them a great weight in the negotiations of the Climate Convention. Being more directly interested in the success of the meeting, they were more willing to make concessions regarding developed countries position. Thus, they played an intermediate role between developed nations and the majority of the underdeveloped world and probably will continue doing so.

"Innocent" countries, that is, those who are not responsible for large emissions, are in a different position. Having little to offer in regard to reduction of emissions contributing to climate change, they do not have strong bargaining powers to request aid. For them, the only argument lies in associating climate change with underdevelopment and environment in general.

Underdevelopment, Poverty, Environment and Climate Change

The position of developing countries in attributing the main responsibility for climate change and its prevention to industrialized nations is objectively well grounded and won final recognition in the Climate Convention. However, their position is clearly weaker in the attempt to condition climate change policies to support general economic development.

Developing countries have argued that environmental problems are caused by poverty and that without economic development they would not be able to contribute to climate change control. This argument is associated with another which has a strong ethical dimension – that "poverty" is the worst kind of pollution. This is the position most strongly defended by the "innocent" countries because, not being responsible for the present climate threat to the Earth's climate, they fear that they would be excluded from eventual help and aid which might be obtained from funds created to support climate change policies. They fear, also, that the establishment of funds for this specific purpose would be made by sacrificing already scarce international resources available for economic development.

From the developed countries' point of view, it seemed clear that underdeveloped countries were raising issues which were more pertinent to general environmental problems than to climate change itself, and to them the arguments therefore seemed misplaced. Since a lot of ill will was created on both sides because of this issue, we must examine it carefully.

It is true that the degree and the extent of poverty in underdeveloped countries, in contrast with wasteful patterns of consumption in developed nations, is a moral and social problem – in a metaphorical sense, then, poverty is certainly a worst kind of "pollution" than that of rivers, oceans and the atmosphere. But it is also true that poverty, in itself, and in a non-metaphorical way, is not a major factor in climate change.

Pollution of streams, the accumulation of urban refuse, and local air pollution originate to some degree from the activities of the poor and their use of urban and rural space. But local pollution of this kind is of little consequence to global climate change, which was the object of the Climate Convention. Air and transboundary air pollution originates mainly in industrial plants and in the consumption patterns of the rich who, whether in developed or underdeveloped countries, drive automobiles, heat or cool their houses and consume the products of modern industry. But the insistence by developing countries in associating development, environmental problems and climate change policies has a common sense objectivity which must be acknowledged.

The greater the economic problems faced by a country, the more difficult it will be for it to consider policies whose objectives and benefits are uncertain and could only be realized on a long-term basis. Such policies could be supported only if associated with more immediate and direct benefits.

Such is the case with climate change policies. For very poor countries and the poorer population in developing countries, it does not seem sensible to spend time, money and effort on the improvement of the Earth's atmosphere if dire poverty and even hunger are their daily lot. The same with governments; pressing problems related to the mere subsistence of the population and the attainment of minimum standards of living surely take precedence over more indirect and distant benefits for "mankind as whole", which is a too abstract proposition to guide immediate action. Without a certain amount of development, it is truly impossible for these countries to commit themselves to climate change policies, unless economic incentives are attached to them.

We must also consider that, from their point of view and for these same reasons, ecological problems other than the climate which may have a more direct impact on their economies should take precedence. This is particularly true for African countries, where destruction of forests, overgrazing and constant cultivation of poor soils, promoted by poverty and the lack of modern agricultural technology – which may have little effect on global climate change – are creating a vicious cycle of increasing poverty and environmental degradation. The same is partly true for the Amazonia, if not as a present reality, at least as a threat for the near future. In this case, the association among development, forest burning and CO_2 emissions is quite clear. But here, also, the national interest in the control of forest burning lies much more in avoiding the real threat of desertification than in its contribution to Global Climate Change.

But the ecological problems of desertification raise another issue to which developing countries dedicated little attention – it is the question of CO_2 sinks and reservoirs. If less destructive patterns of land use are put into practice, they might result in the increase of afforestation and, consequently, of carbon absorption. Modern agriculture technology, which is closely associated with economic development, may be a major target for financed climate change programs which would be of great benefit to the poorer, agricultural countries.

It is difficult to dismiss the arguments presented by underdeveloped countries. It seems reasonable that, having to face immediate and pressing problems which certainly pose a greater threat to the well being of their population than those which might come form a gradual warming of the planet, they should try to condition their support of climate change policies to obtaining aid for economic development. On the other hand, the position of developed countries is also reasonable when they argue that climate convention negotiations did not provide an adequate forum for discussing the problems of underdevelopment.

It is clear that industrialized countries, while willing to pay lip service to the needs of poorer nations, did not accept conditioning climate change international policies and measures to the solution of development problems. They did not agree to increase the level of Official Development Aid (ODA) from their present level of 0.35 percent of their Gross Domestic Product to 0.7 percent, as proposed by the developing countries. This was a main issue of contention. Developed countries did agree, however, to create a fund (GEF – Global Environment Facility) to help developing countries reduce their greenhouse gas emissions.

It took a considerable effort from Brazil, China and India (the "great polluters," who would benefit most from such a fund), to dampen the demands of many small African and Asian underdeveloped countries – the "innocent" ones – so as to come to an agreement with the United

States (the greater opponent to the increase of ODA). Indeed the US only accepted signing the Convention after such an agreement was reached. There was also a considerable amount of pressure excercised on the side by European Community countries on the United States, to try to obtain from that country some concessions regarding developing countries' grievances. Because there were somewhat different positions among both developed and underdeveloped countries a stalemate could be avoided.

Underdevelopment, Technology and the Problems of the Future

Development is intrinsically associated with climate change and global warming in yet another and most important way. If, at present, greenhouse gas emissions from underdeveloped countries are relatively small, prospects for the future are quite different.

The problem is that if poor, agricultural, nonindustrialized countries have neither the technology nor the resources to reduce their greenhouse gas emissions now, they will not be able, for that matter, to avoid its fast increase in the future.

This is a crucial problem, which cannot be avoided in any discussion or agreement regarding climate change, because it is not related only to present conditions, but to the conditioning of the whole future. This question has to do with the path which will be chosen by developing and underdeveloped countries in order to overcome poverty.

Until recently, both developed and underdeveloped countries had a common expectation that economic stability and well being could only be achieved by the latter's following the same industrialization path taken in the past by today's wealthy countries. Yet, given that such industrialization is responsible for today's climate and environment problems, if other nations follow the same path, there will be no possibility of reducing or even stabilizing future emissions of CO_2 and the production of pollutants, even if developing countries were to maintain themselves below the limits which developed countries now find difficult to accept.

No one has had the courage to openly propose arresting development and maintaining developing or poor nations in their present condition. But as far as the environment is concerned, this would surely be the only solution for maintaining greenhouse gas emissions at their present level, unless other paths to economic development could be opened through new technologies. For this reason, technological transfer is certainly a main issue. In this case, there is no difference in bargaining positions between the present "great polluters" (Brazil, China, India, Indonesia) and "innocent" countries. The Convention was sensitive to this problem, clearly stated in Article 4, item 5, as follows:

5. The developed country Parties shall take all practicable steps to promote, facilitate and finance, as appropriate, the transfer of, or access to, environmentally sound technologies and know-how to other Parties, particularly developing country Parties, to enable them to implement the provisions of the Convention. In this process, the developed country Parties shall support the development and enhancement of endogenous capacities and technologies of developing country Parties. Other Parties and organizations in a position to do so may also assist in facilitating the transfer of such technologies.

However, if technology is a central issue, this is a compelling reason for contemplating economic as well as technological aid. Modern, less environmentally destructive production systems are capital intensive and require a qualified labor force. Capital and qualified human resources are extremely scarce in developing countries. Traditional, that is, polluting industrial patterns, though irrational in a medium term or in the long run, may be the only immediate path available if capital is too scarce. Thus, capital as well as technology is also necessary.

Even considering technology alone, we must admit that there are several obstacles to its transfer to developing countries. Technology is imbedded in productive processes and tied up with productive systems. An increase in the technological level of a productive system requires social and cultural conditions, such as improving the educational level of the population – chiefly by increasing primary education, but also by promoting higher education – so as to produce human resources capable of understanding, adapting and utilizing complex technologies.

Development is not attained through the establishment of a few enclaves of modern industry, highly dependent on foreign capital, foreign knowledge and foreign staff. Such enclaves will not prevent the bulk of the economy from operating according to traditional models of destructive industrialization patterns.

Development means global social change. And, if the only hope for climate stabilization and less polluting economic activity lies in technological progress, then the association of environment preservation with social and economic development is crucial indeed.

Energy and Population

The problem of energy production and consumption illustrates very well the point we are trying to make, especially since it lies at the root of economic development processes. It is both one of its conditions and the main indicator of its occurrence. Inefficient production and consumption of energy is clearly associated with emission of greenhouse gases and pollution. Efficiency in this matter depends on advanced technology.

An increase in energy consumption is unavoidable for developing countries if they hope to increase the standards of living of their population. This will inevitably entail an increase in per capita energy consumption, because it will mean a more general access to things like electric light, transport, refrigerators and other domestic electrical equipment. Even if underdeveloped countries maintain their present level of consumption (and poverty), sheer population growth will mean (if present rates of population growth are not altered) a considerable increase in energy consumption per year.

Regardless of what happens, energy consumption in developing countries will surpass that in industrialized countries between the years 2010 and 2025. To understand this figure clearly, one should realize that a population growth of 2 percent per year is common and, in many areas of the world the rate is closer to 3 percent. As small growth in per capita consumption is unavoidable – unless there is a drastic increase in poverty – the developed countries themselves face the very serious problem of being pushed into global environmental changes by events taking place outside their national borders.

This is why multilateral (and sometimes bilateral) cooperation is so important. Developing countries cannot develop in a rational way (where rational means minimizing or reducing greenhouse gas emissions) without outside help. To help them in this direction is in the developed countries' long-term interest.

Moreover, because underdeveloped countries are quite inefficient in their production and consumption of energy, every dollar invested in programs directed at improving their performance in this area will be several times more effective in reducing future greenhouse gas emissions than a dollar invested in the already efficient developed countries. This is a very strong argument in favor of associating global climate policies and aid for technological development.

Of course, controlling population growth would help. But here again a decrease in population growth rates is a phenomenon clearly associated with development, urbanization and general education. All this requires capital investment and technological transfer which must come, at least in part, from developed countries.

Underdevelopment, Political Dependence, Political Instability and the Role of Multilateral Institutions

We have argued up to now as if developing (and, for that matter, even developed) countries were completely sovereign, autonomous and independent nations and that, consequently, the decision to implement or not implement climate change policies rested exclusively on their financial capacity and on the will of their governments. This is hardly the case. Many ecological problems occurring within developing nations involve powerful international interests, which weak local governments are not always able to face. This is certainly true in the case of the "export" of pollutant industries and industrial wastes, wood extraction, and ecologically destructive mining processes. Predatory hunting, fishing and the gathering of native products are sustained by a profitable market within developed countries, outside the control of developing nations' authorities. It is a common fact that companies which carefully observe environmental protection measures in their country of origin completely ignore these procedures when operating in underdeveloped foreign countries, unless governments are strong enough to enforce protection laws – which is seldom the case. For capital hungry underdeveloped nations, it is often impossible to prohibit such activities because they are frequently important sources of much needed foreign currency. Without some kind of international control, it will be impossible to stop or even to decrease such practices. Within this context, multilateral institutions fulfill quite an essential role on the world scene, because they are able to exercise legitimate international pressure on both developed and underdeveloped countries in order to correct this situation.

There is also another problem which must be examined in order to understand the importance of the role of multilateral institutions. Underdevelopment does not mean poverty and low educational level only. It is often associated with political and administrative instability and, more often than we would like to admit, with political corruption and violence.

We must not forget that many underdeveloped countries are artificial nations created by European, American and Soviet imperialistic action. Some of them were constituted by the forceful integration of different ethnic groups and are now being torn apart by ethnic wars. Even when this is not the case, very few of them had enough qualified personnel to manage industrial plants and to operate the complex machinery of a modern state. Civil society is very weak everywhere and unable to develop institutional mechanisms of control over government. In the absence of an organized civil society, government in these countries is often taken over by the few groups which control genuine sources of power: the army, tribal organizations and, most of all, those supported by foreign interests. Thus, political instability is not exclusively an internal problem due only to the intrinsic weakness of underdeveloped countries, but result of power relations vis-a-vis developed countries. It is part of the experience of most of those countries that attempts to oppose foreign powers' interests often result in subtle or open intervention.

Multilateral institutions can play an increasingly important role in the promotion of political stability in underdeveloped countries because they have the legitimacy to act both to prevent foreign intervention and to support legitimate local governments.

We must also consider that, under conditions of political instability, it is often difficult to assure that financial aid will be directed to the attainment of the objectives for which funds were provided. And it must be said again that local rulers are not the only party responsible for the misuse of such funds or, for that matter, for the correct definition of the objectives to be reached. Much of the international support to underdeveloped countries is available in the form of credit-loans which often are only a subsidy given by developed countries to their own industries and products.

In this situation, multilateral institutions may be reliable organizations to assure that:

a) financial aid is directed to support climate change programs or ecological projects which have genuine global interest; and
b) that such programs are feasible and that there are institutions (local or international) capable of efficiently using such funds.

Multilateral Institutions and Non-Governmental Organizations

Considering the problems of political and administrative instability and the weakness of social mechanisms of control over governments in underdeveloped countries, as well as their conflict of interests with developed nations, it is also necessary to evaluate the importance, not only of official Multilateral Institutions, but also of NGOs.

NGOs have become important actors in the ecological struggle. Official multilateral organizations are increasingly listening to their demands and supporting their activities. NGOs can sometimes be irrational in their beliefs, fanatical in the pursuit of their ideals, unrealistic in their demands, and ignorant of social, cultural and political constraints. Yet, many times, they are also the only independent powers in civil society capable of directing policy and controlling the use of funds, because they often are more permanent and better organized institutions than the civil governments. They also have another advantage – often being international themselves, and acting directly through public opinion, they can link developed and underdeveloped countries' social pressures and thus influence both sides in conducting complementary policies.

NGOs are important and even essential actors in environmental questions in yet another important way. Environmental protection and climate change policies seldom bring immediate and easily perceived benefits to individuals or nations. Besides, they often entail the need to change comfortable habits and patterns of consumption. Only ideological commitment provides the driving force capable of motivating individuals, leading governments and mobilizing society to enforce new laws and restrictions. There is thus little hope for success in developing and implementing policies directed at preserving the environment and preventing climate change without the support of NGOs.

Official multilateral institutions are also important in this respect because they are increasingly a forum in which not only governments, but also NGOs, are listened to, and they may influence each other.

Multilateral Institutions and the Scientific Community

The scientific community has been an extremely important actor in environment and climate change polices. It was the scientific community which raised the problems that are now being discussed on the political level, and it is the scientific community which provides the information used in making decisions. Moreover, the scientific community is itself international and can act directly in a multilateral capacity which is, in part, independent of government support. Scientists have the legitimacy to oppose local polices in their own countries, backed by a certain amount of immunity from political persecution which comes from their international support. They are, for these reasons, extremely influential in forming public opinion.

Multilateral institutions have relied heavily on scientific contribution and have provided an adequate forum in which they can exercise international influence.

Limits and Potential of Multilateral Institutions

It should be clear from the preceding observations that multilateral institutions have a very important role in promoting international agreements and the concerted actions which are necessary to control climate change. It is now necessary to analyze some general aspects as well as the limits and potentialities of this role.

Gone are the days when a more naïve generation dreamed of an union of nations which would constitute a World Government above national sovereign states and national interests for the benefit of humankind as a whole. But if this kind of supra-national power proved to be no more than a dream, some of the functions it would perform are essential to world peace and to world economy. Some kind of supra-national authority is proving to be equally essential in facing global problems such as climate change. Multilateral institutions fulfill this role, although in a manner different from that which would characterize a World Government.

Multilateral institutions are not built on the pattern of national sovereign states. They rather resemble a kind of weak feudal monarchy, where the king depends entirely on the loyalty, the interests and the good will of powerful and autonomous barons, some of whom are much more powerful than the king himself. And, since barons are not all equally powerful, some of them carry much more weight than others. The space of manoeuver left to the king lies in the competing and often conflicting interests among barons, both strong and weak, so it is often possible to establish a balance of power in which there is space for concerted action.

In a similar way, multilateral institutions have no real power of their own and have to rely on the exercise of influence. They also are entirely dependent, even economically, of volunteer contributions from member nations and their political space results, in a great measure, from the simultaneous presence of the conflicting interests of, and the need for cooperation among, its members in matters of common concern. Multilateral institutions are essential to the present world order because they provide the space where the delicate engineering of power balance can take place and, thus, cooperation on vital matters can be constructed. But they are paralyzed whenever any of the great powers takes a firm stand against their propositions.

The power differences among developed nations is also relevant to the question of bilateral versus multilateral aid. The largest contributors of funds (like the United States and Japan) certainly prefer bilateral agreements in which they have complete control of whom should be benefited and how. On the other hand, smaller developed countries, whose contributions are relatively large considering their national budgets, but much less significant considering absolute value, tend to prefer to act through a coalition of donors in which the total available funds would be much larger and would allow more ambitious projects.

Opposition and conflict are an essential aspect of multilateral institutions' fields of action. But conflict must be limited or, otherwise, the whole machinery of negotiations cannot be operated. The dissolution of the Soviet Union destroyed one main source of conflict among nations – that between East and West – which was of such an encompassing nature that world cooperation was almost impossible. But the opposition between North and South, or developed and developing countries, although of a different kind, remains as a potential source of conflict and an obstacle to concerted action on a global scale.

The importance and the limitations of multilateral institutions are clearly seen in the case of climate change. Climate change is a global problem – it cannot be prevented without concerted action on the international level. Multilateral cooperation on the scale which is needed for climate change control is very difficult to obtain, considering the extreme diversity among countries and the conflicting interests which set them one against the other on this and other grounds. Without a neutral ground for negotiations, where every nation has a voice and alliances could be made, it would have been impossible to reach minimum agreement.

There is, of course, a diversity of multilateral institutions; general or specific in scope, global or regional in comprehensiveness. As far as climate change is concerned, which is truly a global problem, the most comprehensive of all multilateral institutions, the UN, played the key role. But not in the sense that its initiatives were independent of regional multilateral institutions or as an umbrella under which all of them played a part. Regional organizations like the OECD, G7, G77, and the EC played and will continue to play a decisive role in defining conflicting positions. And, since there is an overlap between membership in these and other multilateral institutions, they also provide the instruments to mediate conflicts and establish power balances. Such was, to a certain extent, the role played by the European community within the OECD in relation to conflict between the United States and underdeveloped countries during the negotiation of the Climate Convention.

The implementation of the Convention commitments will depend heavily on this delicate work of mediating conflicts and negotiating agreements which can only take place within the multitude of multilateral institutions under the umbrella of the UN. And, if the grievances of developing countries, their present limitations, and their increasing importance as sources of greenhouse gas emissions are not seriously considered and taken into account, the objective of controlling climate change cannot be attained.

Final Remarks

We have tried to demonstrate in the analysis of developing countries' positions that climate policies cannot be established independently of national economic and social contexts. Climate change is not an isolated phenomenon but is closely tied to general economic conditions. Development produced it and only further development through new technologies can prevent it.

A climate convention, clearly, cannot deal with all the complex problems derived from the interconnection between global warming and economic patterns of production and consumption. But, unless this interconnection is acknowledged and measures are taken to promote world economic development in a proper direction, it will be impossible to avoid disastrous climate change no matter what the Convention says.

Only under the umbrella of a comprehensive and multilateral institution like the United Nations will it be possible to solve this dilemma. Within its many different programs, the UN will be able to cope in a comprehensive manner with very different but associated problems like literacy, poverty, pollution, political stability and environmental degradation.

The honoring of the Climate Convention commitments will clearly depend on the possibility of coordinating numerous initiatives and programs in the same general direction: promoting self sustainable development in all countries.

All this sounds quite idealistic when we consider the real power game which takes place at the international level. Nonetheless, there has been some progress in the right direction and the Climate Convention is certainly one of them.

I want to thank Dr. José Goldemberg for his helpful comments on many issues in this chapter.

5

Meeting the Demand for Energy in Major Developing Countries in Asia With Lower Emissions: What Support May Be Provided by the New Regime?

Rajendra K. Pachauri

In this chapter some perspectives are presented on how the new regime can help in meeting the energy needs of major developing countries of Asia. The largest developing countries in Asia are China and India, and if one adds to these all the nations of the Indian subcontinent then we would be covering over two billion people. In the last three years the Asian Energy Institute (AEI), which is a networking arrangement of 11 research institutions in Asia, has been working on a study of greenhouse gas emissions and methods of mitigation as well as their costs. This work has been published in two volumes, namely, (i) Global Warming: Mitigation Strategies and Perpsectives from Asia and Brazil, and (ii) Global Warming: Collaborative Study on Strategies to Limit CO_2 Emissions in Asia and Brazil. Some of the key results from these studies are mentioned briefly here. It is relevant to mention that this collaborative research activity also included an institution in Brazil, since it was felt that similarities between large developing countries are strong enough to warrant a comparison across continents.

At the outset it would be useful to look at the global picture with regard to energy consumption as well as certain other relevant indicators in the industrialised versus developing countries. These are presented in Table 5.1.

To explore possibilities in meeting the demand for energy in major developing countries of Asia with lower emissions, we first need to assess how the demand for energy is likely to grow in a few key countries in Asia, if a business-as-usual path is followed. In recent years several

TABLE 5.1: Distribution of Energy and Economic Activity by National Income

	Country grouping by 1988 GNP/person		
	Poorest (< $ 1000)	Intermediate ($ 1000-4000)	Richest (>$ 4000)
Population, billions	3.1 (61%)	0.8 (16%)	1.2 (24%)
GNP, billion 1988 US dollars	1100 (6%)	1500 (8%)	16400 (86%)
"Industrial" energy use, TW	1.6 (14%)	1.1 (10%)	8.5 (76%)
"Traditional" energy use, TW	1.1 (73%)	0.2 (13%)	0.2 (13%)
Total energy use, TW	2.7 (21%)	13 (10%)	8.7 (69%)
Electricity use, trillion kWh/yr	1.1 (10%)	1.1 (10%)	8.4 (80%)
Electric generating capacity, GW	240 (9%)	280 (11%)	2030 (80%)
Refinery capacity, million bbl/day	6 (8%)	1.3 (18%)	55 (74%)
Average GNP/person (1988 US dollars)	350	1900	13700
"Industrial" energy use/pers (watts)	500	1400	7100
"Traditional" energy use/pers (watts)	350	250	200
Electricity use/pers (kWh/yr)	350	1400	7000
Refinery capacity/pers (bbl/yr)	0.7	5.9	16.7

Notes: All figures are for 1988. Parenthetical figures are percentages of category.

Source: Holdren J. P., Pachauri R. K., Proceedings of the International Conference on "An Agenda of Science for Environment and Development into the 21st Century", 1992, 103-118.

projections of energy demand have been made for major countries of Asia. Most notably, a study by ESCAP in 1990 assessed the demand for energy for a select number of countries in Asia and developed projections for the future, employing three different scenarios, namely (1) business-as-usual, (2) emphasis on energy efficiency improvements and (3) fuel switching to produce low carbon dioxide emissions. However, more recently, the World Energy Council (WEC) has prepared forecasts of energy demand for different regions of the world. These were published just before the 1992 World Energy Congress at Madrid. Some of the projections developed in these studies for the WEC are used to illustrate the extent of increase in energy demand that is likely to take place in some countries of the region. As an illustration, Table 5.2 shows the final energy demand for India by sector for different periods extending up to the year 2020.

It can be seen that the total increase in demand in peta joules between 1989 and 2020 is roughly in the order of five to six times the 1989 value. Projections of CO_2 emissions have also been produced for several countries, but there is an obvious variation in these quantities, depending on assumed changes in the mix of fuels that would take place over the next 30 years or so. This rate of growth would undoubtedly increase CO_2

TABLE 5.2: Final Energy Demand For India

S.No	PJ	1980	1989	2000	2020
1	**Final Energy Demand**				
	Agriculture	169	390	615	909 to 1206
	Industry	1588	2146	3960	9940 to 11460
	Transport	649	999	2185	6678 to 8441
	Commercial *				
	Domestic, Govt. &				
	Public Service	396	626	1631	5316 to 8114
	Others		628	827	1638 to 1852
	Non-Energy Use				
	Total	2775	4789	9218	24481 to 31073
2	**Transformation Sector**				
	Conversion losses	756	1829	3299	10307 to 13137
	Energy sector own use	169	303	528	1332 to 1443
	Total	925	2132	3827	11639 to 14580

* included in the domestic etc. sector.

Source: World Energy Council (WEC) Commission, "Energy for Tomorrow's World – the Realities, the Real Options and the Agenda for Achievement", South Asia Regional Report, Madrid, 20-25 September, 1992.

TABLE 5.3: Primary Energy Supply: India

S.No	PJ	1980	1989	2000	2020
1	Coal	2447	3995	7162	19052 to 24686
2	Oil	1423	2479	4548	11554 to 14715
3	Gas	59	430	1040	4650 to 4960
4	Primary Electricity	674	798	1573	3258 to 4528
5	Traditional	4432	4876	5007	2667 to 3638
6	Others	0	3	10	11 to 11
	Total	9035	12581	19340	41192 to 52538

Source: World Energy Council (WEC) Commission, "Energy for Tomorrow's World – the Realities, the Real Options and the Agenda for Achievement", South Asia Regional Report, Madrid, 20-25 September, 1992.

emissions by almost the same order of magnitude as the increase in energy consumption, largely because the indigenous availability of coal in India leads to a preference for coal as the major source of energy in any articulation of future energy policy.

Projections of energy supply for India were also developed in the same WEC study, and while there would be substantial increases in supply of natural gas and primary electricity, the emphasis on coal remains central to energy supply possibilities up to the year 2020. These are shown in Table 5.3.

TABLE 5.4: Energy Intensity: India

S. No	Energy Intensity Data	1980	1990	2000	2020
1	**Primary Energy Supply**				
	GJ per capita	12.43	14.61	18.76	30.07 to 38.19
	GJ per 1987 US $ GDP	0.0046	0.0041	0.0035	0.0023 to 0.0030
2	**Electricity Supply**				
	kWh per capita	175.64	325.66	541.84	1201.52 to 1450.95
	kWh per 1987 US $ GDP	65.34	90.72	100.74	93.46 to 112.87

Source: World Energy Council (WEC) Commission, "Energy for Tomorrow's World – the Realities, the Real Options and the Agenda for Achievement", South Asia Regional Report, Madrid, 20-25 September, 1992.

The result of these demand projections and efforts to meet them through the supply options quantified in Table 5.3 would result in almost a doubling of energy use per capita and roughly a 50% reduction in energy intensity within the Indian economy. This is shown in Table 5.4.

At the same time it must be emphasized that energy use in India as in other developing countries would still remain substantially lower than levels seen today or that can be projected for developed countries in corresponding periods in the future. This subject was studied in some depth in a study carried out jointly by Holdren and Pachauri and presented at the conference on An Agenda of Science for Environment and Development into the 21st Century, in Vienna, 1991. The central thesis of this study indicates that if global emissions levels are to remain below any kind of a ceiling, then the developed countries would necessarily have to bring about major improvements in energy efficiency, leading to a reduction in consumption levels, while the developing countries would consume larger quantities, a reality which must precede the removal of poverty in large regions of the world. Table 5.5 shows some projections that were developed as part of the study, based on a set of scenarios that are characterized by a business-as-usual approach, as well as a more purposeful strategy for improving energy efficiency in the developed and developing countries: incidentally, this would not in any way reduce the level of energy services available for specific end-uses in developed country economies.

Another important implication of total energy consumption, to the extent that this translates into total oil consumption, is the fact that higher levels of demand would result in higher global oil prices. Since a large number of the poorest developing countries are oil importers, the price of oil is an important determinant of the extent to which energy consumption

TABLE 5.5: Conventional Projections for Use of Industrial Energy Forms

	Actual		Projection			
	1980	1990	2000	2010	2020	2030
Population (millions)						
Industrialized	1075	1158	1215	1260	1295	1315
Developing	3310	4085	5000	5900	6750	7575
Energy Use/Person (watts)						
"Business as usual"						
Industrialized	7170	7255	7360	7465	7570	7675
Developing	615	770	965	1205	1500	1880
"Energy Efficient" (Anderson)						
Industrialized	7170	7255	7435	7225	6325	6285
Developing	615	770	950	1340	1720	2300
Total Energy Use (terawatts)						
"Business as usual"						
Industrialized	7.7	8.4	8.9	9.4	9.8	10.1
Developing	2.0	3.2	4.8	7.1	10.1	14.2
World Total	9.7	11.6	13.8	16.5	19.9	24.3
"Energy Efficient" (Anderson)						
Industrialized	7.7	8.4	9.0	9.1	8.2	8.3
Developing	2.9	3.2	4.8	7.9	11.6	17.2
World Total	9.7	11.6	13.8	17.0	19.8	25.7

Note: Business as usual results obtained by extrapolating 1980-90 rates of increase in per capita use of industrial energy forms for industrialised and developing countries. One terawatt = 10^{12}W.

Source: Holdren J. P., Pachauri R.K., Proceedings of the International Conference on "An Agenda of Science for Environment and Development into the 21st Century," 1992, 103-118.

can increase in these nations without causing undue economic stress. If oil prices have to remain relatively steady, then total global consumption cannot increase very rapidly. Even though it is not possible to repeat history with low oil prices, on which economic development in the North was fuelled, it would certainly facilitate growth in the South if oil prices were more or less maintained at current levels.

The purpose of this analysis and the clear conclusion that emerges from it, namely that developed country emissions and energy consumption must come down substantially, is not in any way an attempt to impose equity solutions on the world, but a strategy that emanates from two realities.

1. There is only a limited scope for improvements in energy efficiency in developing countries, including the largest developing countries

of Asia. Hence, any global ceiling has to be driven by a reduction in energy consumption levels in the developed countries.

2. Traditionally, technology transfer has taken place from countries of the north to those of the south. Consequently, methods for energy efficiency improvements and technologies underlying them would largely have to come from the north, which means that efficiency gains would first have to take place in the countries of the north largely for subsequent emulation in the developing countries.

It is important to mention that several recent publications have taken a rather simplistic view of efficiency improvements and computed the potential for efficiency gains in the developing countries based purely on a computation of consumption using estimates for the best technology available. Such computations can prove to be dangerously naive. Technology upgrading is a function, not only of simple technical factors, but a whole range of infrastructural realities, skill levels and institutional issues. In developing countries these factors often inhibit the adoption and dissemination of the best possible technology, and it would be unrealistic to wish them away. In fact, even in the developed countries there are substantial variations in levels of energy efficiency characterized at the one end by the case of Japan, and at the other by the relatively high energy intensity of North America. Efficiency gains in the developing countries would be necessarily gradual.

On another plane, there is the special problem of large rural areas in Asia which are still dependent on substantial quantities of biomass fuels for basic energy uses such as cooking and space heating. For instance, the countries of south Asia consume, in the aggregate, quantities varying from 38% (Pakistan) to 95% (Nepal) of their total energy in the form of traditional fuels. This is brought out in Table 5.6.

TABLE 5.6: Total Energy Consumption (1989) (in Petal Joules)

Countries	Total	Commercial	Traditional
Bangladesh	644	164 (25)[1]	480 (75)
Bhutan	11.2	1.9 (17)	9.4 (83)
Nepal	259	11.9 (5)	247 (95)
Pakistan	1199	730 (61)	469 (38)
Myanmar	219	45 (21)	174 (79)
India	9587	4789 (50)	4798 (50)
Sri Lanka	220	58 (26)	162 (74)

[1] Figures in parenthesis represent percentage shares.

Source: World Energy Council (WEC) Commission, "Energy for Tomorrow's World – the Realities, the Real Options and the Agenda for Achievement", South Asia Regional Report, Madrid, 20-25 September, 1992.

As a strategy the developing countries of Asia could be helped in two specific ways to manage energy demand with low CO_2 emissions. Firstly, solutions have to be found for a shift away from traditional fuels to commercial fuels, which would not only produce lower emissions due to a quantum jump in the efficiency of energy use but a reduction in the drudgery and toil of poor human beings who have to spend a substantial number of hours in the process of collecting biomass fuels and carrying out cooking activities with low levels of thermal efficiency. The second approach would be to ensure that resources and technologies flow from countries of the north to those of the south, ensuring that a reduction in CO_2 emissions takes place wherever it is viable in a global context.

Let us look at the dimensions of the first challenge. Worldwide the quantum of traditional fuels used currently is estimated at 18.726×10^6 TJ. The efficiency of use of this quantity is generally at a level of 8%, yielding, therefore, useful energy output of barely 1.498×10^6 TJ. If this quantity of traditional fuels was to be replaced by conventional fuels, which normally would permit a device efficiency of 50%, this could be achieved through a consumption of 35.78 million tons of oil equivalent (MTOE) of petroleum products annually. In other words, a total consumption of about 1.2% of the total world production of petroleum would be sufficient for reducing 50% of the traditional fuels used throughout the world. At a cost of $20 per barrel of petroleum this would require a total expenditure of around US $5 billion annually. Such higher consumption of petroleum products in several countries would involve import of refined products, particularly kerosene and LPG. In some cases, however, there may be a preference for enhancing indigenous refining, which would involve additional investments in new refinery capacity. The government or the petroleum industry in the country concerned would have to mobilise such investments as required. In addition, consumers would have to make investments in new stoves and connections to use the petroleum product in question for reducing traditional fuel use. Assuming costs that are applicable to India, we have estimated the total level of such private investments of be somewhat in excess of US $2 billion on a global basis. It must be emphasized that such an investment is generally beyond the capacity of the poorest people of the world who would really be the target group for such a program.

These quantities are put forward only as an indicator of the monetary dimensions of this problem. What is far more difficult, and a much more complex undertaking, is the challenge of creating institutions and mechanisms by which a transition could take place from large scale use of biomass to more efficient commercial fuels. However, this is unlikely to happen unless the transition itself is seen as viable by the populations concerned and those who are able to pay for commercial fuels, who have, for generations, been used to collecting non-commercial fuels through the expenditure of time and human effort. If one is to assign an opportunity cost to the time spent in the collection of and cooking with biomass fuels, then we are essentially dealing with an enormous enterprise which deserves the large scale attention and involvement of multilateral organizations, bilateral organizations, intellectuals and researchers, and of course, the governments of the developing countries themselves. The poor in the developing countries are not even part of what is generally considered as the third world. These are citizens of a fourth or fifth world consisting of some two billion people who depend on biomass and traditional fuels.

Biomass consumption remains an important element of energy consumption in the developing countries for several years in the future. It is therefore important to ensure that, while a transition to commercial sources of energy, as described below, is important, it is perhaps even more important to bring about improvements in efficiency in the biomass energy cycle itself. Some of these possibilities would involve the use of very high levels of scientific inputs as, for instance, through the use of modern biotechnology techniques. Research and development on these possibilities has been totally inadequate, if not negligible, and a major step up of efforts is now overdue.

An estimated 500 million women and children in the developing countries spend, on an average, two hours a day collecting biomass resources which adds up to a billion women hours daily. If there were appropriate institutions and opportunities for productive employment for these individuals, even at a modest rate of 20 US cents an hour, the total value of this biomass collection enterprise would be over $30 billion annually. If we add to this the time spent in cooking, then it would be perhaps accurate to double the value of this enterprise. In other words, the fuel collection-cooking system for the poorest in the world can be attributed a value of $60 billion annually. Yet, the attention that this problem area receives, despite the major implications that it has for human welfare and poverty eradication, is meager, to the point of being negligible. What can the world do to solve this problem? Firstly, we need to experiment with projects, with a sense of urgency, in which women could be provided commercial fuels in payment for their time, which can be utilized for productive activities that have a market value. Secondly, the entire range of biomass energy activities requires research and development, so that simple but efficient solutions can be developed even with existing biomass resources. In other words, cooking with biomass can be made far more efficient if science and technology inputs were to focus effectively on this problem area.

If we look at the second range of options where the new regime could help the major developing countries of Asia, then financial and technology transfers are critical to the success of efforts in moderating increases in CO_2 emissions in these countries. A study carried out by the Asian Energy Institute provides estimates of specific options and the extent to which CO_2 emissions could be reduced by adopting them in selected countries of Asia.

To ensure the exercise of efficient options in the optimal use of transfers of finance and technology as envisaged under the Climate Change Convention, it is vitally important to develop local capabilities of various kinds in the developing countries. The mere transplantation of technology from north to south, even with high levels of capital inputs, is unlikely to work. In the past, such efforts involved large scale use of consultants from the north which leaves hardly any capacity or capability to develop this in the recipient countries themselves. The concept of joint implementation which is envisaged under the Climate Change Convention lends greater force to the need for joint research efforts as well. Hence, a new era of joint research projects in a spirit of equal partnership between research institutions of north and south is to accompany other aspects of implementation of the Climate Change Convention.

This Convention now provides the framework for efforts to reduce CO_2 emissions in the years ahead, but the developing countries in general and those in Asia in particular view several elements of the Convention with a sense of fear and scepticism. Given the burdens of poverty and sluggish growth and with the threat of rapidly growing populations, the developing countries of Asia regard the removal of poverty as the most important element in any strategy protecting the global environment.

The Climate Change Convention, which essentially requires the developed countries to provide financial resources and technologies to the developing countries on the basis of "full agreed incremental costs of mitigation of greenhouse gases", involves carrying out a great deal of research and estimation, not only to identify specific opportunities, but to be able to cost them appropriately. Undoubtedly, such efforts at estimation would be most valuable if implemented jointly between countries of the north and south. It would be most appropriate and timely, therefore, for the institutions of the north and south to work together in laying the foundations for equitable and effective implementation of the Climate Change Convention. The 1992 CICERO seminar in Oslo, Norway marked an important step in exploring these possibilities. It is hoped that in the new regime the first measure of support that is provided to the developing countries is a better understanding of the issues involved, and this certainly would require joint efforts between institutes like CICERO and those located in countries of the south.

References

"Economic and Social Commission for Asia and the Pacific (ESCAP)," Symposium on the Climate Change Effects of Increased Fossil Fuel Burning and Energy Policy Implications for the Asia-Pacific, Tokyo, 12-14 December 1990.

Holdren J. P., Pachauri R. K. 1992. Proceedings of the International Conference on *An Agenda of Science for Environment and Development into the 21st Century*," 103-118.

Pachauri, R. K, and Preety Bhandari (eds) 1992. *Asian Energy Institute, Global Warming: Collaborative Study on Strategies to Limit CO_2 Emissions in Asia and Brazil*. New Delhi: Tata McGraw Hill Publishing Co Ltd.

World Energy Council (WEC) Commission. 1992. *"Energy for Tomorrow's World – the Realities, the Real Options and the Agenda for Achievement,"* South Asia Regional Report, Madrid, 20-25 September.

6

Socially Efficient Abatement of Carbon Emissions

William R. Cline

Introduction

This chapter examines prospective global carbon emissions over time under two alternatives: "business as usual," and an "optimal" trajectory in view of the problem of global warming. My cost-benefit analysis of greenhouse policy (Cline, 1992a) and my recent experiments with the dynamic optimization model developed by Nordhaus (Cline, 1992b; Nordhaus, 1992a, 1992b) provide a basis for addressing this issue.

Several major points warrant emphasis at the outset.

- The long-term projections of business-as-usual emissions depend critically on assumptions about growth in population, growth in per capita income, and trends in carbon emissions per unit of GDP.
- With the proper long-term horizon of up to 300 years, global warming could be much more severe than the range usually associated with "benchmark" doubling of carbon dioxide concentrations.
- Economic damages could correspondingly be large.
- However, economic costs of limiting emissions would also be significant, and would tend to occur earlier in the time horizon than the "benefits" of damage thereby avoided.
- Any benefit-cost analysis of greenhouse policy therefore depends importantly on two central elements of the calculation: the "discount rate" for comparing economic values at different points in time; and "risk aversion," or the relative weight assigned to high-damage scenarios.
- Whereas in principle it should be possible to identify a unique "optimal" path for reductions of emissions from their business-as-usual baseline, this calculation is extremely sensitive to the assumptions.

- For policy purposes, it may thus be best to focus on whether or not a broadly agreed "aggressive action" path appears to be socially efficient in the sense that risk-weighted benefits exceed costs of abatement, appropriately discounted.

Baseline Emissions and Warming

Several energy-economic-carbon models now exist that project carbon dioxide emissions under business-as-usual or "baseline" conditions through the year 2100. The discussion that follows begins with these estimates, and then turns to the longer horizon that I consider essential for an appropriate policy analysis of global warming.

Emissions Through 2100. The projection of carbon dioxide emissions typically begins with a forecast of global population and a projection of per capita income. The two estimates combined indicate the growth path for GDP. The corresponding baseline for carbon dioxide emissions then depends on the time path for energy requirements relative to GDP, and the behavior of carbon dioxide emissions per unit of energy over time. Various models have varying degrees of specificity with respect to geographical regions, economic sectors, and technological alternatives for energy production.

The overall patterns of the models tend to be as follows. First, population will tend to reach a plateau somewhere in the latter part of the 21st century. Second, per capita income growth will slow down somewhat from rates of recent decades. Third, there is ongoing technical change that reduces the amount of carbon dioxide per unit of GDP, primarily as the effect of rising energy efficiency.[1] As might be expected, projections can be highly sensitive to assumptions in these various components (Nordhaus and Yohe, 1983). Nonetheless, the principal projections tend to lie within a reasonably coherent range.

Figure 6.1 shows the baselines estimated in five alternative studies, with emissions expressed as billion tons of carbon. The projections here are for fossil fuels, and exclude some 1 billion tons or so of annual emissions from deforestation. The central estimate is provided by the Intergovernmental Panel on Climate Change (IPCC, 1990c) for its business-as-usual case. The IPCC estimate assumes that world population reaches 9.5 billion by 2050 and 10.4 billion by 2100. World GDP grows at 2.1 to 3.3 percent annually in the years 2000-2025, and at 1.3 to 2.6 percent in 2025-2100. At the midpoint, per capita income growth in the second half of the 21st century is thus about one and a half percent annually, and population growth only about 0.2 percent annually. The "moderate" energy efficiency assumption is that the energy/GDP ratio falls by 1 to 1.5 percent annually at the outset, declining to the rate of 0.7 to 1.2 percent annually by 2075-2100.

FIGURE 6.1: Global Carbon Emissions, 2000-2100 (billion tons)

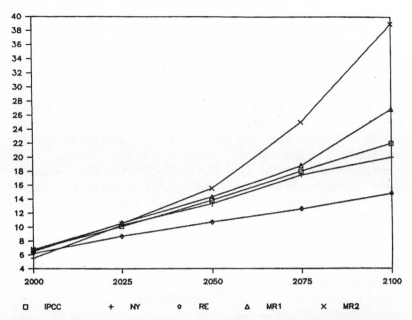

The next three projections shown are from three leading models (Nordhaus-Yohe, NY; Edmonds-Reilly, RE; Manne-Richels, MR1), as reviewed in Cline (1992a, ch. 2). The average for these three models provides the baseline for my own analysis, and places global emissions at approximately 13 billion tons of carbon (GtC) in 2050 and 21 GtC in 2100. However, more recent projections of two of these models have revised baselines upward. Thus, Manne and Richels (1992) have placed emissions by 2100 at 39 GtC (MR2 in figure 6.1), and recent estimates by Edmonds place emissions by that year at 22.6 GtC (Dean and Hoeller, 1992); both estimates are about 50 percent higher than the earlier projections (RE and MR1 in figure 1). The high Manne-Richels projections in particular reflect the increasing carbon intensity of energy as the world shifts toward coal-based synthetics under business as usual in the second half of the 21st century.

As Dean and Hoeller (1992) have shown, the principal difference between the two bracketing projections – the Edmonds-Reilly model on the low side and Manne-Richels on the high side – stems from the assumption about the "autonomous rate of energy efficiency increase," or AEEI, which is set at 1/2 percent annually by Edmonds and Reilly but at 1 percent by Manne and Richels. When the two models apply identical values for AEEI, they obtain similar baselines.

In sum, the most recent projections of the leading models suggest that global carbon emissions are likely to rise from their 1990 level of 5.6 GtC to some 14 GtC by 2050 and reach a range of 20 to 40 GtC by 2100. The IPCC projections are toward the conservative end of this range, suggesting that its global warming projections are more likely to be understated than overstated. The business as usual projections have the sobering implication that annual emissions could more than double by the middle of the next century and rise between fourfold and sevenfold by 2100. Yet, if global warming is to be largely avoided, it will be necessary to reduce emissions from current levels. To do so would require massive proportionate reductions from baseline, on the order of 80 percent or more by late in the coming century.

Baseline Warming. The IPCC has estimated that under business as usual there will be a doubling of carbon equivalent (including other trace gases) above preindustrial levels by the year 2025. After allowance for "ocean thermal lag", as the increased radiation first warms up the ocean, surface temperatures would reach their corresponding equilibrium increase by perhaps two to three decades later. The IPCC reaffirmed the scientific estimates that equilibrium warming from a doubling of carbon dioxide equivalent would be in a range of 1.5°C to 4.5°C. The report set the best guess for this benchmark warming measure, known as the climate sensitivity parameter (Δ), at 2.5°C. The more recent General Circulation Models (GCMs) tend to place Δ higher, at about 3.5 to 4°C (IPCC, 1990a, table 3.2a). However, the best guess was set lower, in part because observed warming of about 0.5°C over the past century is somewhat lower than these models predict, and in part because some scientists expect cloud content change (from water droplets to ice crystals) to increase their reflectivity (albedo) and provide a negative feedback for initial warming.[2]

Recent studies strongly suggest that actual warming has been lower than the GCM predictions largely because of what may be called "masking." One source of masking is from the sulfate aerosols placed in the atmosphere by urban pollution. These particles reflect sunlight directly and also contribute to the formation of low clouds. The overall effect may be to contribute "negative radiative forcing" that is comparable in magnitude to the positive radiative forcing imposed by the greenhouse effect to date (1 to 2 watts per square meter; Charlson et al, 1992).

Masking has also occurred as the result of the placement of smoke aerosols into the atmosphere by the burning of tropical forests (Penner, Dickinson and O'Neill, 1992). These smoke particles also directly reflect sunlight and contribute to low cloud formation, and may have cooling effects comparable to those from sulfate aerosols from urban pollution.

It should thus be no surprise that observed warming has been less than predicted by the GCMs without special incorporation of these two masking influences. Yet the screen provided by urban pollution and bio-mass smoke can hardly continue to grow apace with greenhouse gas concentrations, and is likely to show diminishing returns as a source of cooling. The broad implication is that the proper value for Δ may be closer to the 3.5 to 4°C indicated by the recent GCMs than to the IPCC's best guess of 2.5°C. From this standpoint too, then, IPCC's projections would appear conservative.

The central "business-as-usual" projections by IPCC call for a "com-mitment" (i.e. abstracting from ocean thermal lag) to 2.5°C warming by 2025, and 5.6°C by the year 2100 (IPCC, 1990a, figure 6.11). By implication, with the upper-bound Δ at 4.5°C, the warming commitment could stand as high as 4.5°C by 2025 and 10°C (=5.6x[4.5/2.5]) by 2100. Actual warm-ing would follow the commitment date by perhaps a quarter-century.

Very Long-Term Emissions and Warming

The conventional benchmark for analysis of global warming is the impact of a doubling of carbon dioxide equivalent above pre-industrial levels. This yardstick has been useful for such scientific purposes as comparison among alternative GCMs. However, it is seriously inade-quate for public policy purposes. The greenhouse effect is cumulative and irreversible over a time scale of centuries. The doubling benchmark would be reached already by the year 2025; however, carbon emissions, concentration, and warming can be expected to increase thereafter, as previously discussed.

On the basis of analysis by Sundquist (1990), it would appear that the proper time horizon for policy analysis is some 300 years. By the end of this period, the mixing of carbon dioxide into the deep ocean could begin to reduce atmospheric concentrations. However, until that time, concentrations could reach seven or eight times their pre-industrial levels just for carbon dioxide. There are ample deposits of relatively low-cost fossil fuels (primarily coal) for this outcome to occur without a natur-al choking-off of carbon emissions from rising fuel prices. Other trace gases (primarily methane, the CFC-HCFC group, and nitrous oxide) would contribute further to the greenhouse effect.

To obtain an estimate of potential global warming over a 300-year horizon, it is first necessary to extend the baseline for carbon emissions. I have done so by applying the rates of increase implied by the three prin-cipal models discussed above for the period 2075-2100. As there is little population growth and relatively slow per capita income growth by this time, these projections yield a relatively moderate further increase of emissions, which grow from 20.5 GtC in 2100 to 56 GtC by 2275 (Cline, 1992a, p. 52). If the more recent Manne-Richels projection of 39 GtC by

FIGURE 6.2: Global Carbon Emissions through 2275 (billion tons)

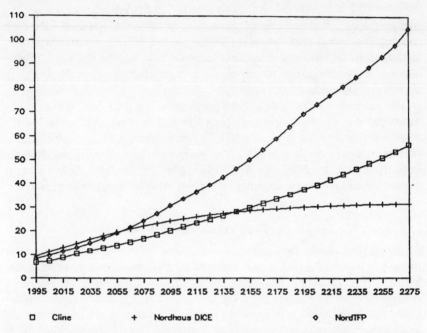

□ Cline   + Nordhaus DICE   ◇ NordTFP

Source: "Optimal Carbon Emissions over Time: Experiments with the Nordhaus DICE model," by William R. Cline. Copyright 1992 by the Institute for International Economics. Reprinted by permission.

2100 are more appropriate, my estimate for 2275 could be seriously understated.

The time path of global warming over the very long-term may be estimated on a basis of these emissions projections and their corresponding implications for atmospheric buildup. These estimates use the IPCC scientific relationships, and take into account the further contribution of other trace gases.[3] The result is that with the "best guess" climate sensitivity parameter, $\Delta = 2.5°C$, actual warming would reach approximately 2.5°C by the year 2050 and 10°C by the year 2275. If the upper-bound sensitivity parameter is considered ($\Delta = 4.5°$), actual global warming would reach 4.5°C by 2050 and 18°C by 2275 (Cline, 1992a, chapter 3).

Nordhaus (1992a, 1992b) has recently reformulated his global warming analysis in a dynamic optimization model, "DICE," (Dynamic Integrated Climate Economy), that also adopts a very long-term horizon of 400 years. Accordingly, he has provided a baseline estimate of carbon emissions over this period. Figure 6.2 reports my very long-term baseline along with the Nordhaus-DICE projection. As can be seen, the two are relatively close through approximately 2150. However, by the year 2275 the Nordhaus emissions reach only about 32 GtC, whereas my estimates place the level at 56 GtC.

The Nordhaus-DICE estimate for 2275 would seem seriously understated, especially in view of the fact that it is lower than the recent Manne-Richels estimate for 2100 (39 GtC; figure 6.1). The central force driving the low Nordhaus estimate is a slowdown in economic growth. In my baseline, global production multiplies 26-fold from 1990 to 2275. This expansion is comprised of a multiple of 2.5 for population and a 10-fold multiple for per capita income. Although the rise in per capita income may seem large, it implies annual per capita growth of only 0.8 percent on average, about half the rate achieved by the United States over the past century (Cline, 1992a, p. 287).

In contrast, in the Nordhaus DICE model global output multiplies only 7.5-fold over the same horizon (Cline, 1992b). As the model assumes approximately the same population growth as I have used, per capita income multiplies only three-fold over the next three centuries in the DICE model, for an average growth rate of only 0.4 percent per year. The main source of this low growth is a steady decay in the rate of technological change. From its initial rate of about 1.5 percent per year, technological change in the production function drops to 0.5 percent annually by the 100th year and 0.18 percent per year by the 200th year. With diminishing returns to rising capital per worker, and with declining technical change, per capita output growth reaches very low levels.

Returning to figure 6.2, if the rate of technical change in production in the DICE model is increased to yield a world GDP path similar to that in my projections, the model generates far higher carbon emissions, reaching over 100 GtC by the end of the horizon (NordTFP). The emissions path is higher than my baseline because the Nordhaus model has a slower rate of efficiency increase in the carbon (energy)/GDP ratio. The principal implication of the alternative path is not that the central expectation should be for much higher emissions than in my baseline, but rather that reasonable changes in the output assumptions can make an enormous difference to the emissions path over such a long horizon.

Primarily because of low global output growth and a low baseline for very long-term carbon emissions, global warming is also relatively low even over the very-long-term in the Nordhaus DICE model.[4] The central estimate for actual warming reaches only 5.25 degrees by 2275 (Cline, 1992b). Considering that the IPPC identifies a higher warming commitment than this amount even by 2100, the warming baseline in the DICE model would appear seriously understated.

The Nordhaus-DICE estimates are preliminary, and are reported here simply because they are the only alternative very long-term estimates for emissions and global warming. However, they do highlight the widening band of uncertainty about economic and emissions futures as the time horizon is stretched to a distance compatible with fossil fuel resources and greenhouse irreversibilities.

One major source of concern about global warming would imply still lower future economic growth than in the Nordhaus DICE model: the fear that future generations will be poorer than the present generation. If that were the case, there would be even greater reason to be concerned about imposing global warming damage on the future generations. It is thus worthwhile considering what may be called a "Malthusian" scenario, in which rampant world population growth exhausts rising production potential and holds per capita income unchanged.

The baseline population projection here assumes that population growth in developing countries falls to zero by 2100. If, instead, it remains stuck at a plateau of only 0.8 percent per year from 2050 onward, then world population reaches 23 billion by 2150 and 48 billion by 2250 (Cline, 1992a, p. 306). The assumption of zero per capita income growth implies low or negligible technological change. A corresponding assumption of zero increase in energy efficiency would place global emissions at 49 GtC by 2275, almost the same as in the base case. Thus, a bleak Malthusian scenario could generate comparable emissions over time despite the absence of rising per capita income, as a consequence of a massive increase in population and minimal gains in energy efficiency.

All of these considerations would seem to imply that a baseline for carbon emissions of some 50 GtC by about 2300 is highly plausible and perhaps understated. Correspondingly, the central estimate for very long-term warming in my analysis, 10oC by 2275, would seem unlikely to be a serious overstatement.

Benefits and Costs of Aggressive Abatement

Many in the public and scientific communities maintain that global warming of 2.5°C by the middle of the next century, let alone 10°C or even 18°C eventually, is unacceptable and must be avoided through reduced carbon emissions. Economic policy makers may be considerably less certain.

My own view is that even an economic approach to the problem leads to the conclusion that aggressive action is likely to be socially efficient. If that is so, then the economic approach reinforces rather than undermines the ecological approach to the problem. The driving forces in the economic analysis are that prospective greenhouse damages are substantial even in the central case and could be extremely high, and that it should be possible to limit the costs of abatement to reasonable levels.

The first step in a benefit-cost analysis of abatement is to identify the size of possible greenhouse damage. I have examined this issue in considerable detail, primarily for the United States (Cline, 1992, chapter 4). Against a 1990-scale US GDP of $5.5 trillion, benchmark-doubling warming of 2.5°C would impose annual damages of $18 billion in agriculture, $11 billion from increased electricity requirements for air conditioning, $7 billion from sea-level rise, $7 billion from decreased water supply, $6 billion in loss of life (valued at lifetime earnings) from heat stress, and further damages from forest loss, species loss, increased urban pollution, increased immigration, ski industry losses, and increased hurricane damage. Possible favorable effects, such as reduced heating requirements, are few and small in value.

The largest damage occurs in agriculture. Some have argued that "carbon fertilization" should offset warming damages. However, the contribution of other trace gases means that the steady-state warming associated with a doubling of carbon equivalent must be evaluated with carbon dioxide concentrations of only one-third above today's levels rather than double. Moreover, there are increasing scientific doubts about the applicability of laboratory results to open field production. In contrast, it has been estimated that the incidence of severe droughts that currently occur with 5 percent frequency would rise to 50 percent under benchmark carbon equivalent doubling. Farm adaptation could provide only partial moderation of losses.

Overall, a "moderate-central" estimate of damages from global warming of 2.5°C would place them at 1 percent of GDP for the United States. If a potentially much larger valuation is placed on species loss, and if account is taken of the side benefit of reducing existing air pollution as the burning of fossil fuels is reduced, the economic value of avoiding global warming under benchmark doubling of carbon equivalent could easily reach 2 percent of GDP.

Because the effects are often non-linear, economic damage for 10°C very long-term warming would be in a corresponding range of 6 to 12 percent of GDP (i.e. six times as high for four times the warming). Under higher damage cases (either $\Delta = 4.5$°C, or a higher degree of non-linearity in the damage function), damages could rise to the range of 20 percent of GDP for very long-term warming. For purposes of policy analysis, the benefits of abatement are the value of the greenhouse damages thereby avoided. It should be possible to avoid the bulk of these damages, although even with aggressive action long-term warming seems likely to reach 2.5°C. My calculations assume that policy action can avoid 80 percent of the damage from greenhouse warming.

On the cost side, there are several energy-economic-carbon models that provide a basis for estimating the economic cost of reducing carbon emissions. Typically they use a "production function" concept in which output depends on labor, capital, and energy. Lower carbon means less energy available and thus less potential production. Most of the models generate estimates in the range of 2 to 3 percent of GDP as the output loss that would have to be paid to reduce carbon emissions by 50 percent from their baseline by, say, 2050. The more detailed models tend to show wider technological alternatives at more distant dates, so that the cost of abatement as a percentage of GDP is lower later in the horizon for an identical percentage cut from baseline.

There is a reassuring adherence of the various model results to what might be expected from first principles. Production theory states that the "elasticity" of output with respect to a factor of production will equal that factor's share in national output. Energy has a share of about 6 to 8 percent of GDP, so the elasticity of output with respect to energy is: $\varepsilon_{QE} = 0.08$ or less. That means that if energy availability is cut by 1 percent, output can be expected to fall by 0.08 percent or less. The models with energy process detail tend to show that the required percentage reduction of energy is about one-half the percentage reduction in carbon. We may call this relationship the energy-carbon elasticity, $\varepsilon_{\varepsilon c} = 0.5$. The output impact of a cut in carbon is then the chained effect of the energy-carbon elasticity and the output-energy elasticity. This approach yields the result that a 50 percent reduction in carbon causes a 2 percent reduction in output: $\%\Delta Q = 50\% \, \varepsilon_{QE} \, \varepsilon_{\varepsilon c} = 50\% \times 0.5 \times 0.08 = 2\%$.

There are two reasons why abatement costs might be even lower. First, there appears to be a near consensus among the engineering-technology community that there exists a significant tranche of low cost or zero cost improvements in energy efficiency. The classic example is the shift to compact fluorescent light bulbs. Economists ask why the energy savings are not generated by the market, but it seems likely that a combination of information costs, financing limitations, distorted pricing incentives for electric utilities, and other influences do lead to this result. On a basis of estimates by the US National Academy of Sciences (1991) and others, it seems likely that a reduction in carbon by about 20 percent could be obtained at zero cost through a move toward best practices in energy use.

Second, forestry offers an initial low-cost way to limit carbon emissions. Reduced deforestation is probably the cheapest way to reduce carbon emissions, at costs somewhere on the order of $10 per ton of carbon. Increased afforestation is also low cost, on the order of $15 to $20 per ton of carbon sequestered. In contrast, major cuts in industrial carbon emissions tend to involve marginal costs (and taxes) on the order of $100 per ton or more.

An international program to plant some 250 million hectares in forest

FIGURE 6.3: Costs and Benefits of Aggressive Abatement of Greenhouse Gas Warming

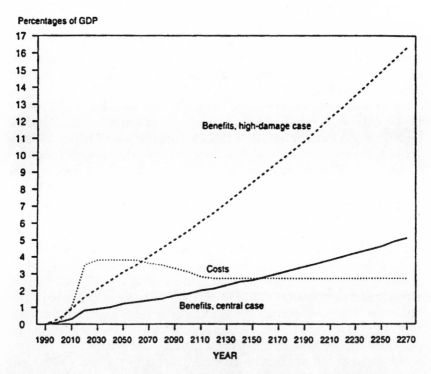

Percentages of GDP

Source: "The Economics of Global Warming" by William R. Cline. Copyright 1992 by the Institute for International Economics. Reprinted by permission.

could sequester about 1.5 GtC annually, although once the forest reached steady-state at maturity, there would be no further sequestration (as carbon released by dying trees would offset that stored by new trees).

The benefits and costs of greenhouse policy may be compared to obtain an overall economic evaluation. Figure 6.3 shows this comparison for an aggressive international program that would permanently limit annual carbon emissions to 4 GtC, or about one-third below current levels and radically lower than the baseline level of emissions that could otherwise be expected late in the next century. This ceiling for emissions is approximately the same as the outcome under the most ambitious abatement scenario considered by the IPCC ("accelerated policies scenario"). The costs here are augmented 20 percent to take account of non-carbon greenhouse gases, and the benefits allow avoidance of only 80 percent of global warming damages.

The abatement costs begin low because of the initial technological "free-lunch" move to best practices and the low-cost forestry options. By about 2015, however, the abatement costs are at their maximum, some 3.5 percent of GDP (based on a cost curve estimated from the various energy-economic-carbon models). Over time the costs decline, because the advent of new technology more than offsets the rising percentage cutback of carbon required. The analysis arbitrarily sets a floor of 2 percent of GDP for industrial carbon abatement; otherwise the cost curve would fall further.

In the "moderate-central" case, shown by the solid line in figure 6.3, the benefits of damage avoided gradually rise to about 1 percent of GDP by 2050 and 5 percent of GDP by 2275. These benefits include effiency gains from using carbon taxes to reduce the excess burden of other taxes (on income, for example), and these gains are relatively important early in the horizon (but of decreasing importance as world GDP rises by the carbon tax revenue stabilizes). In this case, benefits do not exceed abatement costs until about 2150.

The higher damage cases give a very different comparison. As indicated in figure 6.3, benefits of damage avoided are already about 3 percent of GDP by 2050, and rise to 17 percent of GDP by 2275. Nonetheless, even for the high damage cases there is a period early in the horizon when costs of abatement exceed benefits of damage avoided. This "front-loading of costs and back-loading of benefits" is an inherent feature of the greenhouse problem because damage occurs only after a considerable time lag whereas abatement costs are immediate.

As noted at the outset, two crucial elements of the analysis become necessary to compare costs and benefits: discounting over time and incorporation of risk aversion. Figure 6.3 highlights the importance of both. It shows the tendency of benefits to occur later than costs, requiring comparison over time. It also shows the great difference in benefits between a moderate-central damage case and a high-damage case, and thus the importance of taking the risk of high damage into account.[5]

Because of taxes and other capital market distortions, the rate of return on capital is higher than the interest rate received by consumers. Modern discount rate theory (Arrow, et al.; see Cline, 1992a, ch. 6) deals with the divergence by identifying a shadow price on capital. For example, one unit of investment could be worth two units of consumption. With this shadow price in hand, all capital or investment effects are converted to consumption equivalents. Economic evaluation over time is then carried out by applying the "social rate of time preference," SRTP, to the stream of consumption-equivalent effects.

For its part, the SRTP is composed of two elements: a component for "pure time preference," or myopia (π_m); and a component for "utility-based discounting," (π_u), which stems from the anticipation that future per capita consumption will be higher than today's levels and so marginal utility from an additional unit of consumption will be lower in the future than today. Thus, SRTP $= \pi_m + \pi_u$.

Pure time preference exists when consumption is preferred earlier even though future consumption levels are not expected to rise; hence the label "myopic." Considering that average life expectancy in industrial countries is around 75 years, there is not much sense in pure time preference even for an individual. Suppose the individual sets π_m at 3 percent. Then she will be happy to transfer $100 from her consumption at age 75 in order to increase her consumption at age 20 by only $23. That type of approach to personal planning is a recipe for old-age disaster, and is hardly justified by the actuarial tables.

Insofar as we do have evidence on the "pure time preference" applied by the public, it is by no means evident that it is very high. The best indicator of what the public is prepared to pay to have consumption earlier rather than later is probably the real return on treasury bills, a risk-free interest rate. Yet these real returns have been close to zero. Using historical data for the United Kingdom, Scott (1989, p. 231) has estimated the observed rate of pure time preference for households at 1.3 percent. However, for planning society's use of resources over 300 years, any allowance whatsoever for myopic, pure time preference would seem inappropriate. I thus set π_m at zero.

The utility-based component of time preference depends on two elements: the rate of growth of per capita income (g); and the elasticity of marginal utility with respect to consumption (ε). For a broad class of utility functions, it may be shown that the utility-based discount rate should be: $\pi_u = -\theta g$.

In one utility function frequently used by economists, the logarithmic function, the elasticity of marginal utility is unity, so that the utility-based discount rate simply equals the growth rate of per capita income. Scott's empirical estimate of θ for the United Kingdom places its value at -1.5, which happens to coincide with the much earlier estimate made by Fellner (1967).

My analysis applies a value of -1.5 for θ, and assumes long-term per capita income growth on the order of 1 percent (or less, as discussed above), so that $\pi_u = 1.5$ percent per annum. The calculations assume that the share of capital investment in resources displaced by greenhouse policy will be the same as investment's share in the economy generally, or 20 percent. When the corresponding allowance is made for conversion to consumption equivalents (at a shadow price of capital of nearly 2), the overall effect is to raise the time discount rate to about 2 percent.

Even with the SRTP as low as 1.5 percent, the power of compound interest imposes a strong discount of distant future effects. It requires

$20 of avoided damage 200 years from now to make it worth giving up just $1 today. If the SRTP were raised as high as 5 percent, this trade-off would be $17,000 to $1. It is ludicrous to suggest that people will be so much better off 200 years from now that we should impose $17,000 damage on them for each dollar of additional consumption today. There has not been that much progress in the last 200 years.

When this discounting procedure is applied to the "central-moderate" damage and benefits estimates (solid line, figure 3), the resulting benefit-cost ratio for the program of aggressive abatement is only 0.74. That is, discounted benefits of damage avoided cover only about three-fourths of discounted abatement costs. However, policy makers should be risk averse. Accordingly, the analysis incorporates a higher-damage outcome (also shown in figure 6.3) with a higher weight than a lower-damage outcome (not shown in the figure).[6] The result is that the risk-weighted analysis shows a benefit-cost ratio of 1.26. The aggressive program of abatement is thus socially profitable. The benefit-cost ratio can reach much higher levels if even small probabilities of major catastrophes are incorporated.

In sum, a careful economic benefit-cost analysis would appear to support aggressive international action to limit global warming. My policy program suggests that such action be phased in two stages. In the first decade, there would be only moderate carbon taxes (beginning at $5 per ton and rising to $40) to send a signal for technological change. Countries would eliminate existing subsidies for fossil fuels. Intense scientific efforts would be pursued to confirm the severity of prospective global warming, including very-long-term effects. Efforts would begin immediately to reduce deforestation and encourage movement to the energy efficiency frontier.

At the end of the decade, there would be a review of the extent of scientific confirmation of the greenhouse problem. If the problem is essentially confirmed, policy would move to a second and more intense phase, with something like much fuller implementation of the aggressive policy response analyzed above. Thus, higher and internationally coordinated carbon taxes would be put in place. Some portion of revenue would be channeled to developing countries to facilitate their participation in carbon restraints. At a more distant date, it might prove necessary to move to a regime of national carbon quotas, with tradeable permits for efficiency. Such quotas would initially have to include weights based on existing emissions, but would appropriately shift over time to weights reflecting (base-year) population for equity purposes.

The Climate Convention signed at Rio de Janeiro provides a framework for the first stage of this strategy. That convention provides a commitment to begin restraining emissions, and through intentionally ambiguous wording establishes what for practical purposes is a "best efforts" commitment to limit emissions to their 1990 levels by the year 2000. There is no legal obligation to meet this target, but the target is nonetheless implicit. Moreover, the convention provides for interim reviews of progress and scientific understanding, and is thus amenable to the transition to a second phase.

Optimal Emissions Path?

The present policy framework is thus compatible with the broad strategy outlined above. However, some would argue that this strategy is too rigid because it sets a single long-term target (4 GtC annual emissions ceiling) for the second phase. Economists do not like binary solutions (restrain/do not restrain). For example, they analyze market equilibria with supply-demand diagrams that place price on the vertical axis, quantity on the horizontal axis, and identify the optimal production point as the point where a supply curve intersects a demand curve at the unique combination of price and quantity that satisfies both producers and consumers.

A corresponding approach to global warming would seek to identify the optimal time path for carbon emissions, rather than deciding for or against a particular emissions ceiling. In choosing the optimal path, the idea would be that at any point in time there is a social cost curve for abatement (an upward-sloping abatement supply curve) and a corresponding benefits curve (downward-sloping). The abatement cost curve might be low and flat over a certain initial range, but then turn sharply upward. At any point in time the (discounted) marginal benefits of additional abatement should be set equal to marginal costs, to obtain the optimal policy.

Nordhaus (1991, 1992a, 1992b) has been preeminent among economists advocating the identification of an optimal amount of carbon emissions abatement. He has stressed that the initial dose of abatement can be accomplished at very low cost. His choice of parameters and his methodology have led to the conclusion that only a small amount of abatement is economically justified. Thus, in one study he concluded that although CFCs should be largely eliminated, carbon emissions should be cut by only 2 percent from baseline, by imposing a carbon tax no higher than $7.33 per ton (Nordhaus, 1991). For practical purposes, that recommendation was equivalent to doing nothing about the problem. That is, if emissions rise along a path that reaches 49 GtC by 2300 rather than 50 GtC (2 percent reduction), the difference in global warming will be negligible.

In more recent work, Nordhaus (1992a) has concluded that his earlier results were misleading because they used a comparative static approach, whereas the greenhouse problem inherently requires dynamic optimization over an entire time path. He has constructed the DICE model for this purpose (1992b). Nonetheless, his results with the new model are qualitatively similar to his earlier findings that only minimal action is optimal. Specifically, he finds that carbon emissions should be cut by only 10 percent from baseline initially, and only 15 percent by 2100; and that carbon taxes should be held to a range of only $5 to $20 per ton.

The DICE model provides an elegant framework for thinking about the greenhouse problem. It maximizes the discounted sum of utility over time, where per capita utility is a logarithmic function of per capita income, there is discounting for pure time preference, and total utility equals per capita utility multiplied by population. Output is a simple ("Cobb-Douglas") function of labor (population) and capital stock, augmented by a multiplier that captures technological change. For any year, output is divided between consumption and investment. Next year's capital stock subtracts depreciation but adds this year's investment.

The model has carbon emissions per unit of GDP declining over time, but at a decelerating rate. The atmospheric stock of anthropogenic carbon declines at a given rate of decay, but rises by a marginal fraction applied to this year's carbon emissions. Radiative forcing is a logarithmic function of carbon concentration plus an exogenously given contribution from other greenhouse gases (which rises over time at a declining rate). Warming depends on total radiative forcing and the climate sensitivity parameter.

In the model, the decision variables are the percentage cut of carbon emissions from baseline ("z") and the rate of saving over time. Output falls by a cubic function of the variable for carbon reduction (cost function). Output also falls as a function of global warming, but with a quadratic damage function. Optimization then selects the optimal value for "z" at each year over a 400 year horizon.

As suggested above, the parameters Nordhaus applies in his draft versions of DICE (1992a, 1992b) would appear to give a downward bias to prospective baseline emissions, potential long-term warming, and thus optimal reductions in emissions. In addition, his implementation of the model applies a substantial pure rate of time preference (3 percent rather than zero) and does not incorporate the risk of higher-damage cases. Thus, in the two critical dimensions emphasized above – time discounting and risk aversion – his initial application of DICE would appear to generate a downward biased measure of optimal emissions cutback.

Figure 6.4 shows one set of simulations I have conducted with the DICE model (Cline, 1992b). The lowest curve in the field shows a replication of the Nordhaus result: optimal cutbacks begin at about 10 per-

FIGURE 6.4: Rate of Emissions Reduction

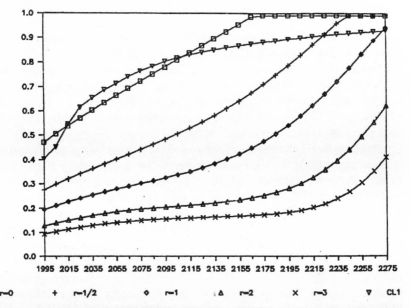

cent, rise to 15 percent by 2100, and (although not reported in Nordhaus 1992b) reach about 40 percent by 2275. The other curves show the optimal results in DICE when all of the Nordhaus assumptions are held unchanged except for the value of the rate of pure time preference. The exception is the curve labeled CL1, which shows the percentage cutback from baseline in my "aggressive action" program.

It is evident that as the rate of pure time preference is reduced, the optimal path for carbon emissions abatement rises and comes closer to my aggressive action program. Because Nordhaus uses a utility function with unitary elasticity of marginal utility (rather than a value of -1.5), his implicit "utility discounting" is at the growth rate of per capita income. As noted above, the Nordhaus long-term per capita growth rate is only 0.4 percent. As a result, his utility discounting is implicitly only 0.4 percent, whereas mine ($\pi_u = -\theta g$) is 1.5 percent. Thus, allowing 1 percent for pure time preference, discounting (π_m) effectively sets Nordhaus' overall social rate of time preference equal to mine (where I have $\pi_m + \pi_u = 0 + 1.5 = 1.5$, and the adjusted respective components in DICE as simulated here are $1.0 + 0.4 = 1.4$).

Along the 1 percent pure time preference path, optimal abatement in figure 4 begins at 20 percent, reaches 30 percent by 2100, and rises to 95 percent by 2275. This path lies considerably above the abatement suggested by Nordhaus; yet it still does not incorporate allowance for risk of higher damage cases, nor does it correct for understatement in the emissions and warming baselines. Other simulations indicate that when economic growth is set equal to the path I have suggested, generating much higher baseline emissions and warming, the optimal cutback path identified by DICE rises sharply. Thus, again with $\pi_m = 1$ percent, optimal cutback begins at 20 percent from baseline, reaches 50 percent by 2100, and 95 percent by 2155.

Unfortunately, as these alternative sets of results suggest, the optimal path of reductions is highly sensitive to the assumptions about pure time preference, emissions baseline, radiative forcing specification, and so forth. In practice, it would seem more promising to seek to derive insights from dynamic optimization, but to use these insights in a qualitative way to tailor the policy response rather than for determining reliable specific timetables for reduction.

One such qualititative insight might be inferred from the shape of the optimal cutback fields in figure 6.4. Especially at the higher pure time preference rates, the optimal paths keep reductions from baseline relatively modest until late in the horizon, and then rise briskly to high percentage cutbacks.[7] The qualitative recommendation would seem to be to take only modest action at first, because discounting makes sharper cuts later in the horizon preferable to more uniform cuts over the horizon.

From one standpoint the percentage cuts are almost inevitably lower at first. With the baseline rising from 6 GtC to 50 GtC, even a flat 4 GtC ceiling means smaller proportionate cuts initially (one-third) than later (e.g. 92 percent cut by the end of the horizon). However, the DICE optimal solutions make the initial cuts even milder and late cuts even steeper, in relative terms. The central reason is that discounting makes it attractive to defer the costs.

The risk of this counsel from the standpoint of political economy is that it legitimates the normal political tendency to procrastinate. The danger is that if today's generation does little, the generation 50 or 100 years from now will be unwilling to take much more intensive measures on grounds that such action is the optimal compensation for lesser action at the outset.

Conclusion

On balance, the most useful approach to economic analysis of the greenhouse problem is probably to consider a plausible path for aggressive policy response and to determine whether it is socially efficient in the sense that appropriately discounted, risk-weighted benefits exceed costs. That is essentially what my analysis of the 4 GtC ceiling does. Further finetuning to attempt to identify the optimal path of emissions cutbacks over time is likely to be even more fraught with sensitivity to key assumptions, so that any particular optimal path must be viewed with considerable caution. At the same time, my experiments with the DICE model do suggest that under more comparable assumptions, a dynamic optimization approach can lead to relatively aggressive action. Extension of these experiments to incorporate the risk of higher-damage cases would shift the field of results further in this direction.

Overall, this analysis suggests support for the premise set forth at the outset: there would appear to be a case for forceful policy response to the problem of global warming on economic grounds, apart from whatever public desire there may be to take action on grounds of more general scientific and ecological concerns. It is time to lay the groundwork for such action, and to plan for more intensive implementation after an initial period for further scientific confirmation.

Notes

1. In some models, the second part of the link from GDP to carbon, the carbon/energy ratio, actually shows an adverse trend rather than technological gains, because of the exhaustion of low-carbon natural gas and intermediate-carbon oil and the shift toward high-carbon coal, directly and in the form of synthetic liquid fuel.

2. In contrast, most GCMs model clouds based on changes in expected relative humidity, which should reduce the amount of low clouds and increase the amount of high clouds. As the former tend to be anti-greenhouse (their reflection of inbound solar radiation is relatively high) and the latter pro-greenhouse (trapping of outbound radiation from the earth is relatively high), this change would cause "positive" or reinforcing feedback of radiative forcing. See Cline (1992a, pp. 23-25).

3. The calculations are essentially as follows. Radiative forcing from carbon is: $R_c = 6.3 \ln (C/C_o)$ wm^{-2}, where C is atmospheric concentration and C_o is the pre-industrial level. Atmospheric carbon concentrations are based on the 1990 atmospheric stock (750 GtC, versus 600 GtC pre-industrial) plus one-half of the cumulative emissions thereafter, under the assumption that the atmospheric retention ratio remains unchanged from recent decades. Total radiative forcing is: $R = 1.4R_c$, based on the ratio identified by the IPCC for 2100. Equilibrium global warming is: $W = R\Delta\beta$ in °C, where Δ is the direct effect and β is the feedback multiplier. There is broad agreement that $\Delta = 0.3$. As a doubling of carbon dioxide equivalent gives $R = 4$ wm^{-2}, we have a direct effect of 1.2°C (=4x0.3). For its part, feedback may be thought of as the multiple of three components: β_w for water vapor (with relatively good agreement that $\beta_w = 1.4$); β_a for "albedo" or reflectivity (e.g. from reduced snow and sea ice), with β_a generally thought to

exceed unity; and β_c for clouds. By far the greatest disagreement is over the value of β_c. With Δ =2.5, the overall feedback parameter is approximately β = 2 ($=2.5°/1.2°$). With β_w = 1.43, the combined feedback multiplier for albedo and clouds would be $\beta_a\beta_c$ = 1.4 (=2/1.43). The projections assume that actual warming lags the equilibrium "commitment" by 25 years, for ocean thermal lag.

4. Low warming in the model is also a consequence of an exogenous additive contribution of radiative forcing from non-carbon sources rather than a multiplicative specification. The radiative contribution from non-carbon sources is low in view of IPCC estimates for the year 2100. See Cline (1992b).

5. There is also a low-damage case, not shown in the figure.

6. The weights are 50 percent for central-moderate, 37.5 percent for high-damage, and 12.5 percent for low-damage cases.

7. Indeed, when DICE is run with my own linear cost curve for abatement, the model generates flip-flop results with no abatement (except the 20 percent "free lunch") in the first few decades and then complete elimination of emissions.

References

Arrow, Bradford, Feldstein, Kury. 1992. *In The Economics of Global Warming* (Washington: Institute for International Economics), W. R. Cline (ed.), chapter 6.

Charlson, R. J. et al, 1992. "Climate Forcing by Anthropogenic Aerosols," *Science*, vol. 255, 24 January, pp. 423-30.

Cline, William R., 1992a. *The Economics of Global Warming* (Washington: Institute for International Economics).

Cline, William R., 1992b. "Optimal Carbon Emissions over Time: Experiments with the Nordhaus DICE Model," (Washington: Institute for International Economics, mimeogr., August).

Dean, Andrew and Peter Hoeller, 1992. "Costs of Reducing CO₂ Emissions: Evidence from Six Global Models," (Paris: OECD, Economics Department Working Papers No. 122).

Fellner, W. 1967. "Operational Utility: The Theoretical Background and a Measurement," in *Ten Economic Studies in the Tradition of Irving Fisher*, pp. 39-74. New York: John Wiley & Sons.

IPCC, 1990a. Intergovernmental Panel on Climate Change, *Scientific Assessment of Climate Change: Report Prepared for IPCC by Working Group I* (New York: World Meterological Organization and United Nations Environment Programme; June).

IPCC, 1990b. Intergovernmental Panel on Climate Change, *Potential Impacts of Climate Change: Report Prepared for IPCC by Working Group II* (New York: World Meterological Organization and United Nations Environment Programme; June).

IPCC, 1990c. Intergovernmental Panel on Climate Change, *Formulation of Response Strategies: Report Prepared for IPCC by Working Group III* (New York: World Meterological Organization and United Nations Environment Programme; June).

Manne, Alan S. and Richard G. Richels, 1992. *Buying Greenhouse Insurance: the Economic Costs of CO₂ Emission Limits* (Cambridge, Mass.: MIT Press)

National Academy of Sciences, 1991. *Policy Implications of Greenhouse Warming* (Washington: National Academy Press).

Nordhaus, William D., 1992a. "Rolling the DICE: An Optimal Transition Path for Controlling Greenhouse Gases." New Haven: Yale University, mimeogr., February.

Nordhaus, William D., 1992b. "The 'DICE' Model: Background and Structure of a Dynamic Integrated Climate Economy Model of the Economics of Global Warming," (New Haven: Yale University, mimeogr., February).

Nordhaus, William D., 1991. "To Slow or Not to Slow: The Economics of the Greenhouse Effect," *The Economic Journal*, 101, no. 6: pp. 920-37.

Nordhaus, William D. and Gary W. Yohe, 1983. "Future Carbon Dioxide Emissions from Fossil Fuels," in National Research Council, *Changing Climate* (Washington: National Academy Press), p. 87-153.

Penner, Joyce E., Robert E. Dickinson, and Christine A. O'Neill, 1992. "Effects of Aerosol from Biomass Burning on the Global Radiation Budget," *Science*, vol. 256, 5 June, pp. 1432-33.

Scott, Maurice F., 1989. *A New View of Economic Growth*. Oxford: Clarendon Press.

Sundquist, Eric T., 1990. "Long-term Aspects of Future Atmosphereic CO2 and Sea-Level Changes," in Roger R. Revelle et al, *Sea-Level Change*, Washington: National Research Council, National Academy Press, pp. 193-207.

7

Progress in the Absence of Substantive Joint Decisions? Notes on the Dynamics of Regime Formation Processes

Arild Underdal

Introduction

The main argument developed in this chapter can be summarized as follows: In evaluating what has been achieved through a regime formation process, such as the UN Conference on Environment and Development (UNCED), we should focus not only on the effectiveness of whatever measures are formally adopted by the conference, but also on the impact of the problem-solving efforts and procedures themselves. The latter may be as important as the former, and including them in the evaluation will often, though definitely not always, serve to shift the balance towards a more positive conclusion.

More specifically, I shall argue that much of the frustration expressed in the public debate about what has been accomplished through UNCED builds upon a misconception of the political process. In particular, it seems to suffer from four biases that appear to be quite common in the evaluation of international cooperation in general: First, UNCED is conceived of essentially or even exclusively as a decision-making process, while in fact there are other important aspects that also merit attention. Thus, the initial stages of the UNCED process can be seen as much as an exercise in search and learning as an effort at negotiating a new regime. Second, evaluation tends to focus essentially on (the absence of) collective decisions. This is clearly a very important dimension, but international problem-solving efforts can also lead to a number of less conspicuous unilateral adjustments, the aggregate impact of which may be as important as the impact of whatever collective measures are adopted.

Third, public debate has so far focussed essentially on governmental positions and actions. Without disputing the significance, or even the necessity, of governmental action, it seems appropriate to remind ourselves that societies can indeed respond through other channels as well – e.g. through shifts in consumer demand or producer supply. Finally, the criticism levied by many environmentalists against the outputs of UNCED has focussed essentially on the lack of specific substantive regulations and the absence of firm commitments to more ambitious emission control targets. By comparison, the institutional and procedural arrangements agreed upon seem to have attracted very little attention. Without questioning the primacy of substance over procedure, I shall argue that institutional arrangements should be seen not only as a poor substitute for "the real thing," but also as a potential instrument for regime development and implementation.

These arguments will be developed in two steps. First, I shall outline very briefly the general assumptions upon which the line of reasoning in this chapter is based. Thereafter I shall explore, in summary fashion, to what extent and how each of these general arguments applies to the UNCED process.[1]

Problem-Solving Is More Than Decision-Making

The ultimate official purpose of the UNCED process is to solve or alleviate a set of more or less interrelated global problems. Effective solutions to these problems require some amount of coordination of action, and joint action requires some kind of collective decision. Thus, there can be no doubt that decision-making is an essential component of problem-solving in this context. The argument to be made here, however, is that conceiving of UNCED and other problem-solving processes essentially or exclusively as a decision "game" can lead us to neglect some quite important aspects of the process. Two such aspects are search and learning.

Conceiving of a regime formation process as a decision "game" leads us to focus on the logic of (collective) choice. Thus, formal decision theory takes as its point of departure the existence of a problem that has been exogenously defined and a fixed set of options for coping with it. By contrast, actors in real-life environmental politics typically have to begin by discovering and diagnosing problems, and then proceed to discover or invent policy options. Actors do not simply "have" a problem; they often have to spend a significant amount of time and effort to determine whether there is indeed a problem that calls for their attention, and – if they determine that such a problem exists – to trace its causes and explore its ramifications. Moreover, as seen from the perspective of policy-makers, options do not simply "exist"; they may have to be searched for and perhaps invented. All this indicates that a major challenge facing governments and other actors, particularly in the initial stages of problem-solving, is to develop a base of consensual knowledge that can serve as a (common) framework for decision-making. By implication, search and learning are integral and essential elements of the problem-solving effort, and in logical sequence the two precede decision-making. Other things being equal, the more novel and/or complex the problem is perceived to be, the more time and effort is likely is to be spent on such "preparatory" activities, and the more critical they tend to be to the success of the overall effort.

Controlling anthropogenic inputs to global climate change is arguably the most complex environmental problem the world has ever faced. The state of "diagnostic" knowledge at the outset could not provide a firm basis for informed regulatory action. It seems fair to say that by the time the climate change issue reached the political agenda, most government officials and segments of the public had, at best, a very rudimentary understanding of the causal mechanisms believed to be operating. Even today, there are many critical gaps in our knowledge that need filling. Yet, over the past few years, a substantial amount of research has been initiated or undertaken to improve our understanding of the inputs of human activities to climate change and of strategies for coping with such change. Moreover, bits and pieces of available knowledge have been disseminated to, and at least to some extent absorbed by, decision-makers as well as "the attentive public". There can be no doubt that a fair amount of learning has indeed occurred on a broad scale and over a relatively short period of time. Most of those who have been actively involved in some part of the process have probably acquired a better understanding of the problem as a consequence of their involvement. Furthermore, by comparative standards, the IPCC efforts at developing a base of consensual knowledge could even be rated remarkably successful so far.

Such progress can not be taken for granted. In this particular case the obstacles to be overcome are certainly far from trivial. First of all, modelling and predicting the impact of human activities on the global climate constitute a formidable challenge to research. So does the translation of estimates of ecological impact into estimates of social costs and benefits, along with predicting the economic, social and political impact of alternative abatement or adaptation strategies. Second, decision-makers in large parts of the world have other and much more urgent and serious problems – be they poverty, internal strife, or external conflict – to attend to, and with only very weak governance capacities to cope with them. Under such circumstances we can easily understand that foreign assistance as well as persuasion will be needed to induce and enable governments to allocate time and effort to what appears to be a relatively remote and diffuse threat. Seen in this perspective, the amount of awareness-raising and involvement already achieved may actually not be all that much below the best that one could sensibly hope to accomplish in such a relatively short time span.

I know of no way to bypass such a phase of problem identification and exploration. Any attempt at short-cutting the process by moving straight into a stage of decision-making would almost certainly have failed, and probably even back-fired.[2] Many students of negotiation argue that actors tend to spend too little time on identifying and diagnosing problems and exploring potential solutions, and that it is useful to separate efforts at inventing and exploring options from those of bargaining and choice (see e.g. Fisher & Ury 1983:59f; Susskind & Ozawa 1992:150). Ideas usually need time and attention to mature, and it is a well-known fact of politics that "premature" launching can easily spoil a good solution. Search and learning are therefore important activities, and a fair amount of progress in awareness-raising and in building a base of consensual knowledge about the causes and ramifications of the problem seems to be a necessary condition for successful negotiations. One important implication for political entrepreneurship is that it is important to design effective procedures for accomplishing this task. One of the intriguing elements of such procedural engineering is to strike an optimal balance between protecting the autonomy and integrity of the scientific process on the one hand, and responding to the "needs" of decision-makers and "brokering" findings from scientific research into premises for substantive decisions on the other (see e.g. Miles 1989:49f).

In building a base of consensual knowledge significant progress has been made through the IPCC and other activities related to the UNCED process. However, consensual knowledge by no means amounts to a sufficient condition for the establishment of an effective regime; aggregating diverging preferences into joint decisions about substantive action is the really hard part. On that score, progress has so far been modest, and the political malignancy of the problem suggests that the prospects of major break-throughs in the near future are not very bright.[3] More knowledge and a better understanding of the problem are no panaceas, either; to the extent that new knowledge makes it easier for actors to predict their own gains or losses from alternative policies, they may serve to strain rather than facilitate negotiations. Seen from the perspective of a political entrepreneur, the existence of a "veil of ignorance" may be a blessing (see e.g. Underdal 1980:24f; Young 1989:16f).

Collective Action Is Only Part of the Overall Response

Global climate change is definitely a collective problem. It is also clearly a politically malign problem, in that it is characterized by: (a) severe incongruities between the cost/benefit calculations of individual actors and that of the "world society" taken as a whole, and (b) great asymmetries in emission volumes and rates, and (consequently) also abatement and damage costs. Even the largest countries harvest only a modest fraction of any impact that their own emissions of "greenhouse gases" may have on the globe's climate system. Also, measures to control such emissions would have equally pervasive external effects. Moreover, since most of the world's economies are to some degree open and hence locked in relationships of competition, the indirect costs (stemming from of a weakening of one's competitive edge) of imposing unilateral emission control measures on one's own industries are likely to be substantial, and might in some cases even dwarf the direct costs of shifting to other sources of energy or more environmentally benign technologies.

The political structure of the problem clearly suggests that, in international negotiations, governments can be expected to pursue, at best, strategies of conditional cooperation, i.e. offer to contribute to solutions if, and only if, others come forth with a certain minimum of contributions in return.[4] The major challenge to the negotiators is to determine a set of mutually acceptable exchange rates for contributions. This challenge is significantly exacerbated by the existence of severe asymmetries among participating countries with regard to, inter alia, contributions to creating the problem in the first place, the relative costs of abatement, and, to some extent, the vulnerability to damage. Other things being equal, the greater such asymmetries are, the more divergent the principles of "fairness" and "equity" invoked tend to be.[5]

Fortunately, however, there is another side to the coin. Even in the case of a "malign" collective problem, governments often also respond through unilateral adjustments. The climate change issue provides interesting illustrations. For one thing, research undertaken in response to the threat has identified a number of no regret options, such as measures for improving energy efficiency. Admittedly, obstacles such as scarcity of investment capital will probably prevent many of these options from being pursued, particularly in Third World countries. Yet, the fact remains that the search for policy measures to cope with the threat of climate change has helped us discover, or called attention to, a number of adjustments that seem to make good sense even in the absence of any link to global climate change. Similarly, the general increase in attention to environmental problems seems to have induced a number of experts and decision-makers to explore the feasibility of adjusting tax systems so as to shift some of the burden towards environmentally harmful activities. In this respect the climate change issue is only one part of a larger complex, but since it has attracted a fair amount of public attention, at least in some societies, it probably has added something to the general case for "greener" fiscal policies. Last, but not least, policy-making processes tend to generate their own stakes (cf. Underdal 1992), some of which are likely to be detrimental, while others can provide positive incentives. I shall explore the logic of process-generated stakes somewhat further in a later section. It may therefore be sufficient here to point out that some of the unilateral emission control "targets" that have been announced by governments and parliaments seem to be heavily influenced by a desire to "perform well" in the eyes of concerned publics. Although some governments may very well fail to meet their own targets, this set of commitments, and the efforts made to implement them, still count for more of the aggregate policy response to the risk of unintended climate change than any collective decisions made so far.

There are at least two interesting implications of this observation. One is that, in certain circumstances, some unilateral action can indeed be expected even in the case of truly global environmental problems. The other is that unilateral and collective action can be linked in important ways, inter alia, so that international negotiations tend to focus attention on what might be called the problem-solving performance of participating governments, thereby generating (additional) incentives to present a good record.

Societies Respond Not Only Through Their Governments

Despite some innovative moves to involve NGOs, and despite a substantial amount of activities on their part, UNCED must, like most other attempts at negotiating international environmental regimes, be considered essentially an inter-governmental affair. It is hardly conceivable that an effective response to the risk of global climate change or loss of biodiversity could be developed in the absence of (inter-)governmental action. Yet, it is important to keep in mind that societies also have other channels for responding to environmental problems, and that adjustments made by non-governmental actors may – at least in some circumstances – be an important element of the overall response. To substantiate this general argument, let me briefly look at two critical subsets of societal responses; viz. shifts in consumer demand and producer behavior.

A number of surveys have indicated a rather strong "willingness to pay" for environmental quality, even in relatively poor countries. Now, we may confidently predict that there will be a significant discrepancy between the professed willingness to pay in principle and actual shifts in consumer demand. Significant shifts in consumer demand can be expected only when (a) one particular "culprit" has been identified, (b) the costs of shifting to some other product are perceived to be truly marginal, and (c) a major educational campaign has been undertaken to drive home these points. Thus, to the average American or European consumer, boycotting whale meat or Norwegian cod is certainly a minor (perhaps even a negative) sacrifice, and so was the shift towards non-phosphatic detergents, once adequate alternatives became available. Unfortunately, the climate change problem does not meet the former two requirements; rather than one culprit there are multiple sources of "greenhouse gas" emissions, and some of these are inextricably linked to essential activities in (industrial) society. Under these conditions, the best that we can realistically hope for in the absence of public policy directives or incentives are truly marginal adjustments in consumer behavior, such as minor shifts towards public transport or more fuel-efficient cars, limited to particular segments of society.

In this particular case, the prospects for adjustments on the producer side may be slightly better. There are several social mechanisms that may induce producers to move. One is simply the educational impact of knowledge dissemination and public discussion. New information may lead corporate leaders to discover new no-regret options, and more generally may serve to increase their interest in the environmental impact of the activities of their companies. To induce changes that entail real costs, however, a change in economic incentives may be required.[6] Policy processes may indeed affect the incentives of societal actors. Thus, the fact that a certain problem has become the subject of public attention and (inter)governmental policy-making efforts, implies a certain probability (perceived by many as a risk) that new regulations will be imposed, affecting the market for the kind of products presently produced (or the costs of continuing with present production or abatement technology).[7] If the risk of new regulations is perceived to be significant, and the impact of such a policy change upon future markets is also believed to be large enough to make a real difference to the firm, a producer will have significant incentives to try to adjust. Moreover, producers may find themselves locked in a competitive game where successful adjustment by one or some producers can significantly increase the probability that new regulations will be imposed upon others as well. New regulations often provide important strategic advantages to those producers that are (most) capable of adjusting, and these producers may – for very good reasons – even turn into active and influential proponents of stricter rules or higher "penalties".[8] The role of Du Pont in the process of controlling emissions of CFCs is an interesting case in point (see e.g. Benedick 1991). If the most "capable" producer(s) also happen(s) to be based in a major market (such as the US), a potent alliance may be forged between environmental NGOs, one or more major producers, and a government wielding a substantial amount of bargaining power vis-a-vis foreign governments and producers.

Again, the structure of the climate change problem is such that this mechanism is likely to be weaker than in, for example, the case of controlling emissions of CFCs. For one thing, the risk that governments will impose stringent regulations appears to be lower. Moreover, even if governments did act to increase, e.g., the price of energy generated from fossil fuels, producers may have good reasons to believe that demand elasticity will by and large be relatively low, at least in the short run. Finally, forging powerful "tripartite" alliances is likely to be harder in the "greenhouse gas" case than in that of CFCs; I suspect that in important areas such as that of energy efficiency, the correlation between technological advancement and size of domestic market will be lower.

Even though the case of climate change may be one where the overall conditions for societal responses may be unfavorable by comparative standards, it seems appropriate to remind ourselves that societies can, and, in fact, do respond to environmental stress also through non-governmental channels. As indicated above, however, some of the adjustments that one might expect from non-governmental actors are to a large extent responses to (anticipated outputs of) the problem-solving process occurring at the (inter)governmental level. Thus, a relationship of mutual reinforcement seems to exist between governmental and non-governmental action. Other things being equal, the stronger the public concern about environmental damage, the stronger incentives governments have to act. Conversely, the more likely that governments will undertake some regulatory action, the stronger the incentives on the part of the societal actors affected to try to adjust in time. And to the extent that they succeed in adjusting better than their competitors, they may in turn acquire significant incentives to push for regulatory measures.

Institutions Matter, Even If They Are Mere Arenas

The analysis above points to a somewhat paradoxical conclusion: In the area of controlling anthropogenic sources of "greenhouse gases", UNCED has so far achieved very little in terms of collective decisions on substantive regulation. In fact, substantive collective action is clearly the area where its record is the least impressive. At the same time, the kinds of progress that are outlined above seem to owe much to the UNCED process; arguably, they are to a large extent spin-offs or side-effects that would not have materialized in its absence. Taken together, these observations suggest two propositions that – in view of the public debate following the "Earth Summit" – may be less trivial than they might seem: First, institutions (here: intergovernmental organizations and conferences) can make a significant difference, also in terms of inducing unilateral moves and societal responses to global problems. Second, building institutional capacity for governance can indeed be a sensible strategy, particularly in the initial stages of problem-solving efforts. It is interesting to note that Weale (1992:14f) found that the first major steps towards upgrading their commitments to environmental quality taken by the European states that he examined, very much focused on organizational arrangements. If building institutions was the preferred starting point for developing domestic environmental policies, such a strategy may be all the more appropriate at the international level, where the existing institutional capacity for problem-solving is generally much weaker. The rationale behind such a strategy is premised on two critical but fairly robust assumptions: first, that institutions can be effective instruments for problem-solving;[9] and, second, that it will most often be easier to design politically feasible solutions to meta-problems (such as determining how to proceed) than to rally sufficient support behind any specific strategy of substantive action.

Let me now try to elaborate these arguments by exploring somewhat more specifically how international institutions might influence actor behavior and outcomes of problem-solving processes.

Intergovernmental organizations can serve at least two main functions in international environmental management; that of being an arena for exchange of information, discussion and decision-making, and that of being an actor in the policy-making or the policy implementation process. All IGOs and conferences serve as arenas, but only a subset (to which UNCED does not belong) can also qualify as significant actors in their own right.

An arena regulates the access of actors to problems (and vice versa), and the access of problems to formal occasions for decisions (cf Cohen, March & Olsen 1976). Moreover, it defines the official purpose and specifies the rules, location and timing of decision "games". In addition, it may develop a distinct informal "culture", into which participants are more or less effectively socialized. To qualify as actor, an organization must (in addition) provide independent inputs into the policy process, or somehow amplify the outputs of that process. To meet these requirements an organization must have a minimum of internal coherence, autonomy, resources, and also in fact engage in some relevant external activities. Without a certain minimum of internal coherence, it would not constitute a unit. Without some autonomy in relation to its members and other actors, it would be a mere puppet, the inputs of which should be attributed to its masters rather than to the organization itself. A minimum of political resources is required to become recognized as a relevant (potential) actor in the process. Finally, by definition a real actor must somehow act, i.e. undertake some activity in the "game" in focus.

Arenas differ in terms of, inter alia, rules of access, decision rules and rules of procedure. For example, membership is in some cases restricted to countries that satisfy certain criteria (as is the case with, inter alia, the Antarctic Treaty System). Other organizations (e.g. the International Whaling Commission) are open to any state that cares to submit a formal application and pay its membership fee.[10] The rules of access to particular (executive) bodies within the organization will often be more "restrictive" than criteria for membership of the organization as such. Consensus is the decision rule most frequently subscribed to in IGOs, but a number of organizations have some kind of provisions for decision-making by voting (usually requiring qualified majority). Finally, we know that procedural arrangements often differ in several respects, e.g., with regard to differentiation into subprocesses (committee work vs plenary sessions), and the role of the committee or conference chairs in producing "(single) negotiating texts". One major research question suggested by these observations becomes: To what extent and how do different rules and arrangements affect the capabilities of organizations-as-arenas to fulfill certain critical functions in the decision-making process? This includes providing actors with incentives to adopt and pursue a "constructive", problem-solving approach; providing procedural opportunities for transcending initial constraints (e.g. by coupling or decoupling issues); and institutional capacity for integrating or aggregating actor preferences.

Organizations – and even specific bodies within organizations – also differ in terms of actor capacity. For example, the Commission of the European Community clearly has a much greater capacity to serve as a political actor than the secretariat of the IWC. As indicated above, actor capacity can provisionally be conceived of as a function of internal coherence, the scope and depth of external autonomy (notably via a vis member states), and the amount of political resources at the organization's own disposal. At a more operational level, this leads to research questions such as: (a) Which (kinds of) functions do IGOs fulfill in different regimes? (b) What institutional arrangements does the organization have for performing these functions, and how much resources does it have at its disposal? (c) What is the relationship between the actor capacity and actual functions of an IGO on the one hand and its problem-solving "success" on the other?

Note that the distinction between organizations-as-arenas and organizations-as-actors does not imply a ranking in terms of importance. Institutions can shape outcomes as much by coupling actors, problems and occasions for decision, and by determining the rules of the game, as by entering the game as more or less significant actors. Arenas are important in their own right and for different reasons. In the familiar terms of systems analysis, we may say that actors influence outcomes by providing inputs (of various kinds) and by amplifying outputs, while arenas are important because they provide opportunities for focusing attention on problems, for articulating interests and beliefs, and for aggregating preferences, and also because they provide a set of rules regulating the game itself. It hardly makes much sense to consider one of these functions as inherently "more important" than another. By implication, the assumption that seems to be implicit in much so-called "realist" thinking – that IGOs are important only to the extent that they have "strong" (supranational) decision rules or a significant actor capability – seems to me to be based on an overly simplistic conception of the dynamics of political processes. In the present international political system most IGOs seem to merit attention at least as much for other reasons.

More specifically, as arenas, organizations and conferences can affect actor behavior and outcomes of collective problem-solving efforts in several ways. First of all, by providing an opportunity for discussion and decision an international conference may have an impact upon the political agenda of governments. The mere fact that a conference is about to take place can help focus attention on a particular problem and thus serve as a stimulus for policy-making at the national level. UNCED seems to have served this agenda-setting function quite effectively. Governments typically respond to "upgraded" issues by demanding more expert advice, and by supplying more resources to the institutions "in charge". Thus, getting an issue on to the political agenda may be a key to obtaining access to the policy-making process and also to new resources. The preparatory stages of international conferences seem, in fact, to provide a "golden opportunity" for environmental agencies, research institutes, and non-governmental groups to enhance their own roles and put their imprint on national policy positions. The problem is still defined largely as an environmental issue, and the scene is still largely "theirs". In both respects things are likely to change as the policy process moves from the initial stages of problem diagnosis and formulation of general policy principles to those of adopting and implementing specific measures.[11]

Second, as pointed out above, conferences and other policy processes tend to generate their own stakes for the participants involved. In distributive bargaining such stakes will often have a detrimental impact on the process, in the sense of adding an extra premium to non-cooperative behavior. Thus, in negotiations about arms control and disarmament there is a very real risk that tactical considerations will lead one or more of the parties to acquire more weapons – to be used primarily as "bargaining chips". Moreover, it is a well-known fact that (public) commitments to a certain position tend to increase the political costs of giving up that position, at least for a certain time. Negotiation processes may, however, also generate "positive" stakes. In general, policy-making processes provide incentives to perform "well" in the eyes of domestic clients as well as other participants. To the extent that "performing well" means advocating or adopting "progressive" environmental positions – and, at least in the initial stages, people strongly concerned with environmental values are likely to make up a disproportionally large fraction of the "attentive public"[12] – conferences like UNCED will provide at least some governments with incentives to go "further" than they would otherwise have done.

Now, such process-generated incentives can not be expected to induce a radical shift in policy; what we are talking about are clearly marginal adjustments. Moreover, it is important to remind ourselves that these kinds of incentives are temporary in character; they tend to fade away with the conference spotlights. Yet, Elliot L. Richardson (1992:175) is clearly right in suggesting that "The attention...focused on the government's response would generate pressure to raise its level", at least under certain circumstances. Moreover, he is also right in suggesting that such incentive-generating mechanisms can be deliberately used for entrepreneurial purposes. One interesting device for generating "positive" incentives is the institution of environmental performance review. The skillful use of performance review procedures can generate significant incentives not only to polish one's image but also to go for real improvements. And one attractive feature of this particular device is that it can be used even in the absence of any substantive agreement on abatement targets or emission control strategies (see Stokke 1992).

Third, institutions-as-arenas can foster informal interpersonal networks (e.g. "epistemic communities"), and facilitate the development of mutual confidence and a sense of common purpose. Benedick's (1991; chpt. 13) account of the London negotiations to strengthen the ozone regime is just one vivid illustration that working together over an extended period of time can foster a sense of common purpose and a strong will to "succeed", and can have important side-effects on interpersonal relationships.[13] Some animosities and tensions must be expected, but the net balance is likely to be positive.

Fourth, international institutions can be important as frameworks for learning. Working together in the context of an international organization or conference is often an exercise in (mutual) education – not only about the objective character of the problem (e.g. the science of climate change), but equally about the concerns and problems that others are facing.

Fifth, problem-solving efforts may be perceived as a signal or "warning" about future action. The fact that a certain problem is subject to policy-making efforts may affect expectations of governments and societies. As indicated above, actors can be expected to respond also to an anticipated or even a possible event, provided that (a) the probability of the event itself is above a certain critical minimum (and just meeting the trivial requirement that p>0 will in most cases not be enough), and (b) the event is believed to make a "significant" difference, should it occur.

Note that all the five categories of impact that are listed above can occur even in the absence of any substantive collective decision (although the latter requires a certain probability that substantive action will be taken). The general implication of the argument is simple, but may bear repeating: even a conference that accomplishes nothing in terms of substantive joint decision may – through various mechanisms, some of which are sketched above – lead to positive change. From this conclusion follows another: institution-building merits serious attention as a strategy for problem-solving, particularly in the initial stages. In a very crude outline, such a strategy could include the following steps:

First, the establishment of a suitable arena for problem-solving. The establishment of such an arena implies firm procedural commitments to meet in order to examine the state of the problem, to report national problem-solving efforts, and preferably also to evaluate progress. It furthermore implies commitment to some official purpose and agreement on a set of "rules of play". The UNCED process has made some useful accomplishments on this score.

Second, building some actor capability into a relevant organization could further enhance problem-solving capacity. As indicated above, an international organization can "act" by providing distinctive inputs and/or by amplifying outputs. In the early stages the input functions will be the more important. Suffice it here to observe that a number of organizations or bodies (e.g. ICES and IPCC) have contributed substantially to the production and dissemination of (consensual) knowledge. Scott (1976) argued that this is the most important function that an IGO can possibly fulfill in relation to international negotiation processes. Although perhaps an overstatement, the basic point is well taken: an independent capability for developing a base of consensual knowledge can be a very significant instrument for regime creation. Such a capability can be acquired without building a large supranational bureaucracy; properly staffed and funded, a fairly small core capable of cooperating constructively with national experts and perhaps a few NGOs may be sufficient. Furthermore, problem-solving can benefit substantially from the presence of one or more capable political entrepreneur(s) (cf. Young 1991). Properly arranged and staffed, international secretariats and conference or committee chairpersons can provide important entrepreneurial services, such as establishing informal channels of communication and designing solutions that are politically feasible and at the same time substantively as effective as "circumstances permit". Political entrepreneurship is very much associated with leading figures; for example, much of the entrepreneurial achievements of UNEP can, it seems, be attributed to the skills and political clout of its Secretary General, Mr. Tolba. In terms of building institutional capacity for entrepreneurial action, the achievements of the UNCED process seem so far to be rather meager. But then it should be added that conference presidents or secretariats never will be in a position to provide sufficient leadership all by themselves; for this purpose, they critically depend upon the active support of the delegations and governments of important countries.

Once some substantive joint decision has been made, organizations can amplify outputs by, inter alia, monitoring compliance, carrying out its own projects, or allocating resources to help members implement decisions or achieve agreed goals "themselves". Even as a "passive" arena an organization can add something to a collective decision, notably its "seal of legitimacy". Thus, a resolution passed by the UN Security Council would probably in most countries command somewhat higher political status than a set of unilateral but substantively equivalent declarations given outside the framework of the UN. Interestingly, the fact that a (temporary) ban on commercial whaling had been formally adopted by the appropriate intergovernmental authority (IWC) seems to have served, in internal discussions, as the principal argument against the decision taken by the Norwegian government to permit some commercial catch from one particular stock in 1993.[14]

Third, for a strategy of regime formation through the building of arenas and organizational actor capabilities, it is absolutely essential that there be some demand for substantive action. Thus, the relationship between institution-building and substantive action is symbiotic: without demand for substantive deeds there will be no political energy to drive the process or the institutional "machinery" that it may produce; but without a suitable institutional framework the push for action is most likely to be frustrated and may even backfire.

Admittedly, the kind of incremental strategy sketched in crude form above can by no means be considered heroic. Yet, the more politically malign the problem, the greater seems to be the advantages of incrementalism over its bold alternatives, at least in the initial stages of regime formation processes.[15]

Notes

1. The label "the UNCED process" is used here in a wide sense, denoting not only the Conference itself but also the preparatory work at all levels as well as follow-up activities. On the other hand, I refer mainly to the global climate change issue, leaving out other items on the UNCED agenda.

2. A case can be made for strategies such as learning by trial-and-error. A trial-and-error strategy would, of course, require substantive action. It seems to me, though, that such a strategy can be applied only when the parties involved have at least reached a firm common understanding that they do indeed have a collective problem. Much of the efforts made in the early stages of the UNCED process have been devoted to the development of such a common platform.

3. The (neo)functionalist proposition that initial success tends to increase the demand for further cooperation – through spill-over and other mechanisms – should be tempered with the reminder that if we adhere to their advice to solve the politically "benign" problems first, the partners will face increasingly "malign" issues as they proceed (the notions of "benign and "malign" problems are developed in Underdal 1987).

4. Most governments seem to differentiate their demands upon others. For example, the EC seems to insist upon positive contributions from the US and Japan, but is prepared to accept small and poor Third World countries as "free riders". One general implication of this observation is that only some countries qualify as pivotal members of a climate change regime (on the notion of pivotal actors see Underdal 1992:225f).

5. In the climate change negotiations, at least two basic criteria for the distribution of the costs of remedial action seem to be widely if not universely accepted, even by those who stand to lose from their application. One may be called the principle of guilt, that costs should be distributed in proportion to one's contribution to causing the problem. The other is the principle of capacity, implying that costs should be distributed so that the marginal loss of welfare suffered in paying these costs is equalized (cf progressive taxation). Even though we are still far from general agreement on their relative weight and practical implications, it seems fair to say that UNCED achieved more in terms of agreement on the basic principles of cost-sharing than in terms of commitment to specific "deeds".

6. Incentives need not be material or economic. Another social mechanism that may be operating is that of "social stigmatization". At least in Germany, some industrial leaders have privately confessed concern about their kids becoming subject to unfavorable remarks from classmates about the environmental damage caused by their father's (or mother's) company or plant (private communication from Dr. Helmut Weidner, WZB). So far, these seem to be single instances, and the climate change problem again seems to be too complex for stigmatization to work well (among other things, there are too many "culprits"). Moreover, the moral implications of actively pursuing a strategy of social stigmatization seem dubious; after all, the children must be presumed innocent. But parents are usually concerned about the welfare of their kids, and to the extent that kids are in fact blamed for their parents' association with environmentally harmful activities, the latter may certainly respond (and not only by calling in a public relations consultant to help polish the image of the company).

7. A shift in consumer demand may have similar effects.

8. This observation hints at one of the basic principles of political "engineering": There is nothing as politically attractive as a measure that offers private profits to organized interests and at the same time caters to public morale. Thus, from a political feasibility perspective, foreign aid tied to purchases from domestic industries seems to be a far better proposition than domestic restructuring to open one's home market to Third World producers. The misfortune of environmental policy is that it is relatively poor in projects meeting this dual requirement. Examples can be found, however: the plans to help Russia reduce harmful emissions from industrial plants and energy facilities on the Kola peninsula through state-funded deliveries of modern production or abatement technologies produced by Finnish or Norwegian companies is just one case in point.

9. To repeat, I do not suggest that institutional arrangements can serve as a substitute for substantive action. Moreover, implicit in my analysis is the assumption that they can accomplish the kind of spin-offs examined above only to the extent that they are infused with political energy geared to the search for substantive solutions.

10. In fact, failure to pay membership fees seems not to imply automatic exclusion. This applies not only to the IWC, but to several functional organizations and even the UN as well.

11. The policy process generated by global environmental problems can be expected typically to follow a characteristic pattern: In the initial stages, attention is focused on diagnosing the problem and developing some basic and general principles for responding. The problem itself seems truly global in its ramifications, and the policy options discussed at this stage tend to be rather vague, implying also that the domestic distribution of costs and benefits will be more or less indeterminate. So far, the problem can be claimed to fall largely within the domain of the environmental branch of government, with some inputs from other "holistic" ministries such as those of foreign affairs and finance. However, since environmental damage typically occurs as unintended side-effects of other perfectly legitimate activities – such as industrial production or transportation of people and goods – environmental policy cannot simply be added to other policy commitments; to succeed it somehow has to penetrate the activities that cause harm in the first place. This implies that as policy ideas become further developed and specified, it will become increasingly clear that many of them can have substantial consequences for particular sectors of the economy or segments of society. On their own merits, such specific measures would normally fall largely within the domain of "sectoral" ministries or agencies. Moreover, they are likely to activate the "clients" concerned and their organizations. Over time, this tends to lead to a redefinition of the issue and a "take-over" by specialized sector agencies, and the two are likely to reinforce each other. What started out as a somewhat diffuse "grand design" to cope with some global problem tends to become decomposed and redefined into "micro-issues" of industrial policy, etc. And seen from the latter perspective most of the policy options in question will most likely appear less attractive or urgent than they did when seen as integrated parts of a more comprehensive environmental program. As a consequence, there is a serious risk of ending up with what might be called a vertical disintegration

of policy (Underdal 1979:7), i.e. a state of affairs where the aggregate thrust of "micro-decisions" deviates more or less significantly from what policy doctrines and principles would lead us to expect. Note that the logic of politics is such that neither is likely to reflect accurately the "true" preference structure of government (if such a thing exists) – general doctrines and principles are likely to be "biased" in favor of environmental values, while specific "micro-decisions" tend to be "biased" in favor of protecting the activities causing environmental damage.

12. Such a bias is likely to exist also in the conference itself. For example, thanks to the synergistic impact of the official purpose of the conference or organization and the selection of participants on the basis of formal roles and reputed expertise, the average participant of UNCED is likely to be "greener" than the average participant of GATT.

13. Peter Haas (1992:32) even refers to "epistemic communities" as "tacit alliances", and James Sebenius (1992:354) concurs in describing such a community as a "de facto, cross-cutting, natural coalition of 'believers'."

14. The fact that the IWC decision to ban commercial whaling was dismissed by some as the result of "illegitimate" pressure by environmental groups and media "coercing" the IWC to depart from its official purpose and management procedures, indicates that such a seal of legimacy is attributed to an organization only as long as it is perceived to abide by its own (constitutive) rules. This may be interpreted as just another piece of evidence that "institutions matter". I am grateful to Helge Hveem for calling my attention to this aspect of the controversy over whaling.

15. I readily admit that this proposition is questionable. There is no doubt that political drama and exogenous shocks can, in certain circumstances, provide very important stimuli for action. However, the suggestion by Bergesen (Aftenposten (a) 13/05/1992) and others that Norway and other "progressive" countries could have "kicked" the process further by walking out or refusing to accept the somewhat tepid commitments to substantive action that the US and other laggards were prepared to make strikes me as implausible. My reading of the reports that came out of Rio suggests that the state of the "patient" simply was not such that the kind of cure prescribed by Mr. Bergesen could have worked. For one thing, there probably was not sufficient concern on the part of some pivotal actors that such a move could have generated the kind of "shock" or impetus required to spur demand for, or support of, a more ambitious program. Moreover, the refusal of Norway and a few other small, "like-minded" countries could hardly have been a sufficiently strong stimulus.

References

Benedick, Richard E. 1991. *Ozone Diplomacy*. Cambridge, MA: Harvard University Press.

Bergesen , Helge O. 1992. Interview. *Aftenposten*, (a), 13 May 1992.

Cohen, Michael D.; March, James G.; and Olsen, Johan P. 1976. "People, Problems, Solutions and the Ambiguity of Relevance". In James G. March & Johan P. Olsen, *Ambiguity and Choice in Organizations*. Oslo: Universitetsforlaget.

Fisher, Roger & Ury, William. 1983. *Getting to Yes*. Harmondsworth: Penguin Books. (First published by Houghton Mifflin in 1981.)

Haas, Peter M. 1992. "Introduction: epistemic communities and international policy coordination". *International Ogranization*, Vol 46, No 1, pp. 1-35.

Miles, Edward L. 1989. "Scientific and technological knowledge and international cooperation in resource management". In Steinar Andresen & Willy Østreng (eds), *International Resource Management*. London: Belhaven Press.

Richardson, Elliot L. 1992. "Climate Change: Problems of Law-Making". In Andrew Hurrell & Benedict Kingsbury (eds), *The International Politics of the Environment*. Oxford: Oxford University Press, 1992.

Scott, Anthony 1976. "Transfrontier pollution: are new institutions necessary?" In OECD: *Economics of Transfrontier Pollution*. Paris: OECD.

Sebenius, James K. 1992. "Challenging conventional explanations of international cooperation: negotiation analysis and the case of epistemic communities". *International Organization*, Vol 46, No 1, pp. 323-365.

Stokke, Olav S. 1992. "Environmental performance review: concept and design". In Erik Lykke (ed), *Achieving Environmental Goals. The Concept and Practice of Environmental Performance Review*. London: Belhaven Press.

Susskind, Lawrence & Ozawa, Connie 1992. "Negotiating More Effective International Environmental Agreements". In Andrew Hurrell & Benedict Kingsbury (eds), *The International Politics of the Environment*. Oxford: Oxford University Press.

Underdal, Arild. 1979. "Issues Determine Politics Determine Policies". *Cooperation and Conflict*, Vol 14, No 1, pp.1-9.

Underdal, Arild. 1980. *The Politics of International Fisheries Management*. Oslo: Universitetsforlaget.

Underdal, Arild. 1987. "International Cooperation: Transforming 'Needs' into 'Deeds'." *Journal of Peace Research*, Vol 24, No 2, pp. 167-183.

Underdal, Arild. 1992. "Designing Politically Feasible Solutions". In Raino Malnes & Arild Underdal (eds), *Rationality and Institutions*. Oslo: Scandinavian University Press.

Weale, Albert. 1992. *The New Politics of Pollution*. Manchester: Manchester University Press.

Young, Oran R. 1989. "Bargaining, Entrepreneurship, and International Politics: Escaping the Dead Hand of Nash and Zeuthen". Paper prepared for the ISA annual convention, London, 28 March – 1 April, 1989.

Young, Oran R. 1991. "Political leadership and regime formation". *International Organization*, Vol 45, No 3, pp. 281-308.

I gratefully acknowledge useful comments from Helge Hveem, Rune Sørensen, and participants at the CICERO seminar to an earlier draft.

8

The Legal Status of the Commitments in the Convention on Climate Change and the Need for Future Revisions

Rudolf Dolzer

A Framework Convention

The Climate Convention signed in Rio de Janeiro in 1992 by states carries the official title "United Nations Framework Convention on Climate Change". Recent international environmental practice was followed and a two-step "framework approach" was adopted in which a first agreement, the "Framework Convention", in principle, regulates the broader issues, leaving details and specifics to subsequent agreements, so-called protocols. Both agreements are intended to share the same legal binding quality; the legal rules eventually applicable to concrete issues are determined in a complementary manner by both the framework convention and the protocol.

Nothing within such an approach prevents the parties from drafting the framework convention in a manner which for specific areas results in binding concrete legal obligations on the basis of the framework convention alone. The negotiating parties are free to allocate duties and responsibilities to the first or the second level. While the framework agreement, as such, fully obliges all parties, the precise undertaking assumed by States parties can only be established by way of careful examination of the wording of the text. In a framework treaty, more than in a regular convention, the parties may choose to include language which is general and leaves broad room for interpretation and for further development. If so, these areas have to be specified in subsequent protocols.

The question may well be asked why the two-step framework approach has been adopted in the sphere of international environmental relations. This approach is not entirely unknown in other areas of law, but it is dominant only in environmental matters. Thus, it was chosen in the context of regulating transboundary pollution in the Agreement reached in 1979 and in controlling the CFC's in the rules of Vienna and Montreal.

There is nothing inherent in environmental matters which would prescribe or force the framework approach. Typically, it is the complexity of the subject area and the effort to separate in the negotiations general and specific issues which speak in favor of the two-step approach. In terms of negotiating tactics it allows states to break down and to narrow existing differences of interest among each other. One beneficial side effect is the possibility to establish certain basic obligations, often of a preparatory nature, before reaching agreement on the more difficult details. As regards entry into force of the framework convention, no rule exists which would indicate that negotiations on the protocol have to wait for the framework to become binding. Legally, it would be possible to arrange the rules so as to separate the negotiations but to couple the entry into force of convention and of protocols. The Climate Convention sets forth that it will become binding 90 days after the deposit of the fiftieth instrument of ratification.

It would not be surprising if this point is reached during the coming 24 months. The United States has already ratified. The government in Bonn has decided to initiate the proceedings for ratification in November 1992. It may well be that Pacific Islands states and other AOSIS states will ratify before long; together, more than 30 states belong to this group.

Principles in the Convention

Article 3 of the Convention contains several principles which are intended to guide the parties in their actions to achieve the common objective and to implement the rules agreed upon.

Before addressing the specific obligations agreed upon in the Convention, it is worth considering the substance and the significance of these principles. In general international treaty practice, it is rare to include general principles in the operative text of a treaty. Normally, the text is drafted without expressing underlying principles, or the principles are spelled out in the preamble, with little weight for the substantive obligations undertaken by the parties. The special practice in the case of the Climate Convention is related to the wishes of the developing states. They felt it was in their interest to spell out explicitly the grounds for the difference in the obligations of developed and developing states, both with regard to the origins of the problem and the resulting consequences. Given the framework function of the Convention, this effort was not all academic. The groundwork was laid for the differentiation of obligations to be allocated later on in the protocols. Keywords are "equity," "common but differentiated responsibilities," and a reference to "respective capabilities." In order to avoid any misunderstanding as to the practical implication of these abstract concepts, it was then added that the developed states "should take the lead in combating climate change and the adverse consequences" (Art. 3 sec. 1). Along the same lines, a section is added to highlight the specific needs and circumstances of developing states (Art. 3 sec. 2). The precautionary principle in the absence of full scientific certainty was added at the request of European and low-lying Pacific island countries, an unusual coalition whose viewpoints were several times identical. Cost-effectiveness is also spelled out as a principle. Forest states in particular may benefit from the clause which requires that all relevant sources, sinks and reservoirs will have to be taken into account. The right to promote sustainable development is mentioned alongside the need to integrate national development programmes with climate change policies. Developing states also insisted upon the link between climate change, the economic welfare of developing states and an open international economic system; thus, they would be in a position to argue that they are not required to fulfill their duties in case industrialized states refuse to cooperate in efforts toward a more open system of economic relations. With respect to trade issues, arbitrary or unjustifiable discrimination or disguised restriction are prohibited. Here, the contours of general guidelines for future discussions on the relationship between the requirements of free trade and environmental protection emerge.

On the negative side of the list of principles, it is remarkable that neither the principle of sovereignty as such nor that of permanent sovereignty over natural resources appear. Against the background of the constant emphasis of these concepts in political declarations by developing states, the absence of these phrases in the operative text of the convention must be seen as a tribute to a cooperative spirit in the negotiations; the principles were relegated into the preamble, thus enjoying almost no weight in the application and interpretation of the Convention. It must also be noted that the polluter pays principle as such has not been included; this text of the Convention addresses primirlarily preventive approaches, includes reference to adaptation measures, but is silent on the issue of reparation and damage.

Lack of the Timetables and Targets

The main criticism and disappointment of the Convention on the part of the European states concerned the lack of specific obligations to reduce emissions within a certain period of time. The effort to agree upon such a far-reaching scheme was blocked mainly by the United States, but also by Japan. What is less known is that France was also skeptical about setting precise dates. Public statements on the Convention and its evaluation mainly focused on this issue of "timetables and targets". Less attention was paid to the fact that, within this area, as within others, the negotiations aimed at compromises rather than black and white solutions and that they were successful in this endeavour.

The compromise between targets and timetables on the one hand and total avoidance of all reference to quantification of desirable emissions within given periods on the other hand consisted in

a) agreement upon the objective of the convention
b) a weak reference to the desirability to the return by the year 2000 to earlier emission levels
c) a general statement on the aim to return to 1990 levels
d) a time-table for further negotiations on protocols to specify and to implement the convention.

All these elements of compromise belonged to the most difficult part of the negotiations and were drafted with utmost care. Thus, their wording and substance deserve special attention, and not only because that these provisions went largely unnoticed, overshadowed by the paramount issue of specific timetables and targets.

Objective of the Convention

Article 2 spells out the "ultimate objective" of the agreement, ultimate being in contrast, presumably, to "immediate" or "direct". The objective is to stabilize emission of greenhouse gases at a level which would prevent dangerous interference with the climate system; stabilization of the climate itself is not mentioned as a goal. A time horizon to achieve this aim is spelled out in Art. 2 as well. This, however, is a result-oriented description rather than a date-oriented concept. Three desirable results set the framework for the implementation of the aim to stabilize:

- natural adaptation of ecosystems to climate change and concern for continued food production
- and sufficient flexibility for economic development in a sustainable manner.

It will not be overlooked that these three provisions also bear the stamp of a compromise formula with potential elements of tension among these aims.

Considering the general approach of the agreement on objectives in

Art. 2 and its evaluation, two opposing points of view may be set forth. It is possible to argue that the objective is phrased in such general terms that it does not amount to more than a banality which leaves states entirely uncommitted. The alternative perspective would point out that the objective as phrased in the Convention sets a precise framework which could not be more explicit in policy terms and leaves it to a current knowledge and its future evolution to spell out the practical significance and implications of the policy thus established.

As long as there is no consensus among states with regard to the dangers emanating from greenhouse gas emissions, the objective thus formulated will not lead to any precise consequence. Beyond the difficulty to predict emissions of CO_2 cycle are not well known, and the contribution of greenhouse gases other than CO_2 will have to be studied far more intensely. Also, the effects of aerosols on climate are largely unknown, and the general issue of climate sensitivity will also have to be the subject of more research. In this context, it must be noted hat the uncertainties relate not only to climate change, but also to changes in regional precipitation patterns and the development of sea level rise. Thus, I would hesitate to interpret the clause with unqualified enthusiasm as containing a progressive concept. Nevertheless, my view is that the clause is a most useful one inasmuch as it indirectly points to the urgent necessity on the international level for a common scientific evaluation of the greenhouse problem and, equally important, clearly requires that the scientific basis for the policy discussion must take center stage rather than ideological or economic concepts, thus underlining the necessity for strictly scientific guidance unimpaired by extraneous considerations.

In my view, this approach is useful and appropriate because it directs the policy discussion to those issues which indeed must drive the policy considerations. In discussing future protocols, whoever departs from the scientific starting point places themself outside of the Convention. Of course, this does not settle the matter against the background of scientific uncertainty. Coupled, however, with the precautionary principle agreed upon in Art. 3, this formulation of the objective forms a strong argumentative basis for those who call for negotiations of stringent protocols. While not free from uncertainty in terms of practical application, the objective laid down in Art. 2 clearly represents a major step in the direction of a policy approach which appropriately responds to the dangers of climate change.

Time Aspects of Climate Policy

It is obvious that timing of measures to reduce emissions may turn out to be essential in the formulation of global climate policy. While scientific aspects may call for quick action, issues of economic efficiency also have a bearing upon timing. The reasons for the central controversy of timetables are related to the combined aspects of science and economics.

In its final form, the Convention contains the following time-related elements beyond those implicit in the definition of the objective:

a) In regard to obligation by the developed States, the parties recognize that the return by the year 2000 to earlier levels of anthropogenic emissions of carbon dioxide and other greenhouse gases would contribute to the modification of longer-term trends which are consistent with the objective of the convention (Art. 4 sec. 2 (a)).
b) In the same part regulating duties of developed states, it is spelled out indirectly that these states will aim, individually or jointly, to their 1990 emission levels by the end of the present decade (Art. 4 sec. 2 (b)).
c) The parties have agreed to start negotiations for a protocol within one year after entry into force of the Convention.

Taken together, these time-related clauses require that the parties to the Convention will soon start negotiations "with the aim" of reducing emission levels in 2000 to the level of 1990. The question is raised as to which obligations arise from a commitment which does not prescribe such a reduction, but sets forth the will of the parties "to aim" at such a reduction. In legal terminology, such an arrangement would presumably be called *a pactum de contrahendo*, an obligation to negotiate to reach an agreed common goal, the goal being the return, by the year 2000, to 1990 emission levels. Consequently, any statement that a party, either a developed or a developing state, which ratifies the Framework Convention remains entirely free of obligations would be far more misleading than correct. For the negotiations on the protocols the goal is set, and only the methods and means to reach that goal form the subject of negotiations. A party which would argue that it feels stabilization should, in principle, be reached only in the year 2010 or later, would not fulfill the duties undertaken unless it could demonstrate compelling reasons to depart from the goal agreed upon in the Framework Convention.

The legal situation thus created leaves open the question of whether the same approach chosen in the Framework Convention would also be preferable in the protocols or whether a definite date should be set. Two different philosophies exist in this respect in existing international discussions. One school of thought considers that it is impractical and unfeasible to set a precise date for matters so complex as policies on carbon dioxide. The second approach assumes that the existence of an exact date promotes that spirit of urgency which is necessary to achieve progress in areas so entrenched in different interest groups. I assume that a precise date would lead to prompter action than a general guideline, even though it may not be possible to meet the target at midnight of the foreseen day.

Future Adaptation of Convention

Given existing uncertainties surrounding the climate issue, it was to be expected that the Convention would address the necessity to adapt the Convention to evolving knowledge and circumstances; thus, an evolutionary approach is clearly foreseen (in Art. 7 sec. 2 (a)), and the Conference of Parties will periodically re-examine the obligation of Parties and the institutional arrangements. The Conference will be assisted in this task by the subsidiary body for scientific and technological advice to be established under Art. 9.

The procedure to amend the Convention, established in Art. 15, provides that an effort must be made to reach a consensus; as a last resort, a three-fourths majority vote of the Parties present and voting at a meeting will suffice. Once the amendment is adopted, it will enter into force only if three-fourths of the Party to the Convention has ratified. An Annex to the Convention, being of a scientific, technical, procedural or administrative character, that has been adopted by the same majority will enter into force for all those parties who do not object to it within six months after they have been formally notified of the adoption of the Annex.

Means and Methods of Reduction

Since the beginning of the negotiations in the INC, there was no doubt that the principle of "freedom of means" on the national level would have to be observed; accordingly, it is clear that each state will, in principle, retain the power to decide about the ways and means in which it will fulfill its quantitative obligation to reduce gases. Given the differences in the relevant economic and other sectors at the national levels, any effort to draft a comprehensive global reduction plan including methods for all states would obviously be futile and counterproductive.

Correspondingly, the Framework Convention does not call upon the parties to take any specific action, but instead requires them to adopt plans and programms for reduction which fulfill their quantitative obligation in a manner suitable and acceptable for their national circumstances and priorities. In setting forth those obligations, the Framework

Convention distinguishes between the duties of all parties and those of the developed countries.

All states are required to prepare national inventories for sources and sinks of emissions (Art. 4 sec 1 (a)). Also, they have to formulate and implement national or regional programs containing measures to mitigate climate change by addressing sources and sinks. For developing states this latter obligation forms the core of their substantive duties. It is worthwhile to note the level of generality of the relevant wording. The programs must "contain measures" and "address sources and sinks"; no qualification or quantification was added, and it is clear that no substantial kind and degree of commitment is required by these terms. For some developing states, it may turn out to be important that they have agreed to promote sustainable management of sinks and reservoirs, including biomass, forests, oceans, and other ecosystems. A general obligation also exists for all states, including developing states, to take climate change considerations into account in their social, economic and environmental policies and actions. Legally speaking it is not easy to delineate the contours of such an obligation "to take into account", in any event, climate issues must not be ignored. This duty is, however, further weakened by the qualifying words "to the extent feasible".

In essence, the provisions so far mentioned spell out the substantive obligations of developing states' parties to the Convention: the establishment of inventories, the setting up of programs not in any way specified, and the will to take into account the issue of climate change.

Consistent with the origin of relevant emissions, the Convention sets up much more demanding duties for developed states. They must adopt policies and take measures which demonstrate that they are taking the lead. The teeth in the obligation are formed by the time requirements discussed above. The aim to return individually or jointly to 1990 reductions at the end of the present decade, as required by these clauses, applies to developed states, no comparable aspects of time being included for developing states.

Beyond the more direct obligations to address emissions, it is important to note that all states, including developing states, have to cooperate in preparing for adaptation to the impacts of climate change, and must promote and cooperate in research and in the exchange of relevant information. In the context of education, training and public awareness, the role of non-governmental organizations is specifically mentioned (Art. 4 sec. 1 (i)). Concerning economic and administrative instruments to combat climate change, developed states have the duty to coordinate their approaches, according to the wording of the Convention, only in relation to other developed states.

The Convention allows for individual or joint implementation (Art. 4

sec. (2 (a)). The concept of joint implementation has in no way been spelled out. It seems to allow for a cooperation between states, but also between enterprises. Obviously, such issues as identification, verification and supervision of such projects remain to be considered in detail. Generally, economic instruments, including tax schemes, are not addressed in the Convention.

Categories of Countries

The distinction between developed states and developing states is central to the allocation of responsibilities of the parties to the Convention. However, the Convention goes clearly beyond this primary differentation. In Art. 4 sec. 1, it is generally said that specific national and regional development priorities, objectives and circumstances will have to be taken into account. The implication of this willingness and effort not to treat different situations in an arbitrarily equal manner finds more specific expression in the subsequent part:

- The Convention recognizes differences, among developed states, in their starting points and approaches, in economic structures and resource bases, in the need to maintain growth, in available technology and in other circumstances, concluding that all states have to make an equitable and appropriate contribution (Art. 4 sec. 2 (a)). The vagueness of this wording is quite conspicuous, reflecting difficult compromises. It would not be correct to assume that the Convention has adopted a "grandfathering approach" which accepts that past emissions, or their reduction in the past, will not be considered. The reference to the factors listed above and to the concept of equity clearly indicates a much more multifaceted approach which looks to all relevant circumstances.
- The issue of "graduation", and the change of status of developing countries generally is recognized and addressed in Art. 4 sec. 2 (f). Such a change, however, requires the consent of the state concerned.
- According to Art. 4 sec. 4, the distinction between developed country parties and developing parties is made.
- Special significance was attributed to countries in the process of transition to a market economy. These states are allowed, according to Art. 4 sec. 6, "a certain degree of flexibility". Obviously, this formula reflects a compromise which will need further refinement and definition when more specific responsibilities will be allocated in protocols. In principle, these states are treated as developed states.
- As to developing states, it is set forth, in Art. 4 sec. 8, that their

commitments will have to be seen in terms of their requirements concerning funding, insurance and transfer of technology. Special reference is made in this context to no less than 11 groups of developing states, e.g. small island countries, countries with high urban atmospheric pollution, land-locked countries and the least developed countries. A special paragraph (sec. 10) is devoted to the treatment of states whose economies are vulnerable to the adverse effects of the implementation of climate change measures. Mainly, but not exclusively, oil-producing states have pressed for this clause which gives them a certain flexibility as well. Looking at all these groups of states allowed special treatment in context, and their precise obligations and exemptions remain to be determined; the relevant phrases in the Convention have deliberately been phrased in a vague manner. It is not difficult to predict that the issues raised may turn out to be most difficult and complex in the negotiation of the protocols.

Finances and Technology Transfer

Foreseeably, the developing states focussed their interests in the negotiations on the transfer of financial means and of technology from the industrialised states. They were successful, as mentioned above, in pointing out explicitly in the Convention that their ability to fulfill their duties would depend upon the availability of finances and of technology. Indeed, these issues remained until the very end of the negotiations. Considering the importance of the issues, the substance agreed upon in the text of the Convention remained short, and the matter remains to be further developed in the future. Obviously, the practical significance of the Convention will largely depend upon finances and technology.

As to finances, the Convention requires that "new and additional financial resources" be granted by the developed states (Art. 4, sec. 3), without specifying the precise meaning of these terms; a certain guidance in this respect can be found in Resolutions of the UN General Assembly, referred to in the Preamble of the Convention. Developing states will receive funds covering "the agreed full incremental costs of implementing measures." Again, this wording reflects a carefully drafted compromise, the key words being "agreed full incremental". As the meaning of "full incremental" may in general and in particular instances turn out to be far from clear-cut, the word "agreed" was added. In practice, this will mean that future negotiations will have to define which measures and costs are covered by "full incremental". It also remains to be determined what is meant by the requirement of "adequacy and predictability" in the flow of funds. Rather difficult questions also lie behind the reference in this context to "the importance of appropriate burden sharing among the developed country Parties"; this may refer in particular to national economic factors and/or to the level of greenhouse gas emissions.

On the organisational side, the Convention provides for a Financial Mechanism (Art. 11) which will provide resources on a grant and concessional basis. This mechanism shall have an equitable and balanced representation with a transparent system of governance.

For the time being, the Global Environmental Facility will act as the financial mechanism (Art. 21 sec. 3), and the Conference of Parties will have to decide whether this solution will be the final one (Art. 11 sec. 4). Thus, the GEF is confronted with the task to change its organizational structure along the lines indicated in the Convention. This process is ongoing. It will not be overlooked in this context that financing of tasks within the framework of the Convention must be seen to have a different quality than traditional financing of developmental projects. Within the Convention, developed and developing states have joined together, not on the basis of voluntary one-sided assistance for the benefit of one state, but as sovereign equal partners addressing a common global issue caused mainly by developed states. The demand for equitable and balanced representation forms the corollary to this new situation. It took some time until this context was fully appreciated, but at this point the principle of restructuring as such seems to be out of controversy.

The key provision on transfer of technology says that the developed states "shall take all practical steps to promote, facilitate and finance, as appropriate, the transfer of, or access to, environmentally sound technologies". This wording obviously reflects the compromise between the will of developing states to have free access to relevant technology and the viewpoint to refer strictly to market conditions and the need to protect property rights. Thus, the issue has to some extent only been solved in principle, and further negotiations will have to be held.

One conceivable solution might be tied to the national inventories required by the Parties. It might be possible, once these inventories exist, to establish or to name an authority which would be charged with the identification of technologies suitable to support the efforts of developing states in view of their special situation. Basic issues will be whether the cooperation required will mainly be of a bilateral or of a multilateral nature and whether it will focus on financial support or on technological transfer. Obviously, these approaches are not exclusive and can be made mutually compatible. The Convention itself requests that the development of endogenous capacities and technologies shall be supported. Again, this might involve financial and technological cooperation. The technological issue has also been addressed in Chapter 34 of Agenda 21, also adopted in Rio. The relevant provisions are much more elaborate than those in the Climate Convention. Article 34.4 speaks of the need for favorable access to technology, and it is also pointed out that a large body of useful technological knowledge lies in the public domain, not covered by private proprietary rights. Beyond these clauses, the chapter evades the more difficult issues, as have the negotiations of the Climate Convention.

Reporting System

There is no uniform system in international treaties practice concerning the manner in which the State parties agree to assure compliance with the rules agreed upon. It is quite common not to include any special provision in this respect; the underlying assumption is that the State parties have entered into agreement voluntarily and in good faith, and that this very fact may ensure compliance. As there is no general enforcement mechanism on the international level, states, however, often agree to set up one or the other scheme intended to promote compliance.

In modern multilateral treaties, a frequently adopted approach consists of a reporting system which ensures that states' parties regularly present their measures to comply and to discuss them with other State parties. In the Climate Convention, a scheme of "Communication of Information" related to implementation was adopted. Again, developed states and developing states have different obligations commensurate to the substantive rules. Developing states report for the first time within three years after the entry into force of the Convention, developed states within six months. Developing states may, if they wish, propose projects for financing, together with an estimate of costs and benefits.

Of course, the importance of a reporting system lies less in the communication itself than in the manner in which the report is treated and relevant consequences. In this respect, the text of the Convention does not at all address the issue. However, the Conference of Parties will consider the establishment of a multilateral consultative process for the resolution of questions regarding the implementation of the Convention (Art. 13.). By definition, this procedure, being consultative, will not be binding in a strictly legal sense. Presumably, the communication sent by the party in question will form the basis of such a procedure. One open question, still to be determined, concerns the admissibility of evidence other than the one submitted by the party. Another matter left open concerns the distinction between this form of procedure and the formal system of dispute settlement established in Art. 14.

In my view, these clauses in their sum deserve special attention as they are the ones intended to make up the compliance mechanism. As a practical matter, it must not be forgotten that, indirectly, these provisions must be relied upon with regard to use of the money and technology to be provided by donor countries. Moreover, and perhaps even more importantly, one should remember that in this context, as in any scheme of multilateral obligations, a state may be inclined to attribute less importance to its own efforts; third states aware of, or suspicious of, a lack of compliance may in turn hesitate to fulfill their obligations. It is not difficult to argue that given the special multilateral effort required in the Climate Convention, a stronger scheme of compliance would have been appropriate. However, the issue of sovereignty looms behind these matters from the point of view of developing states, and perhaps also of others. The compromise achieved may turn out to be a viable one, but I would not be surprised if this turns out to be a specially weak spot in the Convention. Contrary to the rules in the ozone regime, the Convention does not provide for special sanctions in cases of non-compliance. Much will depend upon the consultative implementation mechanism still to be negotiated. A sign of hope may also lie in Art. 12 sec. 6 which provides that the Parties may further consider this problem area.

Institutional Aspects

Possible concern about issues of compliance are not eliminated in view of the institutional arrangements of the Convention. The question has been whether, in addition to the Conference of Parties, an efficient, largely independent institution would be created which would monitor scientific developments, contribute to ensure compliance and enjoy sufficient authority to prompt necessary international action.

Before the negotiations in the Convention, a number of possible approaches were discussed. Some assumed, as did I, that a kind of an Environmental Security Council might be created, patterned to some extent after the existing Security Council, and that one of its tasks might be to monitor compliance with global environmental conventions. However, the negotiations in the INC made it clear rather quickly that the time for such an innovation is not ripe. Such a new authority would imply, to some extent, a renunciation of national sovereignty. The reluctance against such a step came from different corners and perspectives. The third world, still preoccupied with a sovereignty free of colonial ties, and also the more powerful states, considered that national freedom of action remains a paramount point of orientation, and the plans for a new institution so far came to nothing. It is often said that a proliferation of international institutions has to be avoided, but the European experience has shown that the effective pursuit of common goals does require institutional backing.

The solution adopted in the Convention is a minimal one; the establishment of a Secretariat (Art. 8). In order to ensure that the Secretariat will not become too powerful, its weak powers are spelled out explicitly in an enumerative manner excluding other powers. The Convention provides explicitly, however, that the Parties will periodically review this matter (Art. 7 sec. 2(a)). According to Art. 9 (a), the Conference of Parties will periodically examine the obligations of the parties in the light of the objective and the available best scientific and technological experience. The first review will take place at the first session of the Conference of Parties, and the parties will then be free to adopt amendments (Art. 4 sec. 2(d)). It is foreseen that such amendments will be reached by way of consensus, but a three-fourths majority vote will also suffice (Art. 15 sec. 3); it is no use to speculate at this point how such a majority might be reached and what implications could arise. Amendments must be proposed six months in advance.

Future Evolution

The future evolution of the Convention will primarily depend upon the scientific assessment of the dangers arising from the greenhouse effect. The Convention provides a framework within which a flexible response is being facilitated and promoted. The philosophy of the Convention is characterized by its blend of a few substantive decisions on the level of principle and the compromise-driven open-endedness in major areas of concern.

Thus, the Convention could turn out to be a failure if the will to push international efforts to combat the greenhouse effect have been exhausted by the INC negotiations. It will not be overlooked that the current financial situations of most developed and developing states are not conducive to strong and prompt international action. Also, it seems that many political short-term problems constantly occupy the attention of policy-makers more than long-term ones surrounded by scientific uncertainty fraught with complex economic and social policy issues.

There are hopeful signs that the entry into force of the Convention after 50 ratifications will not be long delayed and that the process of negotiations of protocols, especially on CO_2, will be started before long. My prediction is, however, that the challenge of negotiating these protocols may well surpass that of negotiating the Framework Convention. Thus, it is urgently necessary to build on the political momentum created by the INC negotiating process itself and the new consciousness apparent during the UNCED meeting. Correspondingly, the gap between the signing of the ratification of the Convention by more than 150 States and the negotiations and entry into force of the Protocols must be filled by the prompt start announced by several states. The early adoption of national programs and of policies, and the translation of the spirit of cooperation expressed in Rio between North and South into financial commitments in the coming months will serve as the yardstick by which to measure, for the time being, the real effect and value of the Convention.

9

Efficient Abatement of
Different Greenhouse Gases

Michael Hoel and Ivar S. A. Isaksen

Abstract

Although CO_2 is the most important greenhouse gas, there are a number of other greenhouse gases which are important for the development of the climate. An efficient climate policy should in principle be related to emissions of all climate gases, weighed together with their impact on the climate. Constructing a climate policy which is efficient across greenhouse gases would in principle be straightforward if the relative impact of different greenhouse gases on the climate was given by some fixed physical coefficients. This is, however, not the case. The most important reason for this is that the relationship between emissions and the development of atmospheric concentration differs significantly across greenhouse gases, since different greenhouse gases have different lifetimes in the atmosphere.

In this chapter we present a model which illustrates the importance of the above issues for the appropriate weights of the different greenhouse gases. Using optimal control theory, the conditions for intertemporally efficient time paths of greenhouse gas emissions are derived. The weights of the different greenhouse gases relative to CO_2 are at any time given by the relative costate variables associated with the differential equations describing the development of the different greenhouse gases. It is shown how these weights depend on assumptions about important characteristics of the future economy. In particular, the discount rate and the assumed growth rate of the economy are important for the weights of the different greenhouse gases. Properties of the function describing the costs of climate change are also important.

Introduction

Although CO_2 is the most important greenhouse gas, there are a number of other greenhouse gases which are important for the development of the climate. An efficient climate policy should in principle be related to emissions of all climate gases, weighed together with their impact on the climate. In particular, if the emission of one ton of a particular greenhouse gas has an impact on the climate which is a times as large as the impact of one ton of CO_2, then 1 ton of this gas should be treated in the same way as a tons of CO_2. This means that the efficient abatement level of this greenhouse gas should be so large that the marginal costs of further emission reductions are a times as large (per ton) as the marginal costs of reducing CO_2 emissions.

Constructing a climate policy which is efficient across greenhouse gases would thus in principle be straightforward *if* the relative impact of different greenhouse gases on the climate was given by some fixed physical coefficients. This is, however, not the case. There are two reasons for this: In the first place, the climate development depends on the development of the radiative forcing, which in turn depends on the development of the atmospheric concentration of all greenhouse gases. At any time, the relationship between atmospheric concentration and radiative forcing is non-linear, i.e. the relationship is more complex than a function from a weighed sum of concentrations to radiative forcing. Secondly, and more important, the relationship between emissions and the development of atmospheric concentration differs significantly across greenhouse gases. The reason for this is that different greenhouse gases have different lifetimes in the atmosphere.

In this paper we present a model which illustrates the importance of the above issues for the appropriate weights of the different greenhouse gases. It is shown how these weights are quite sensitive to assumptions about important characteristics of the future economy. In particular, the discount rate and the assumed growth rate of the economy are important for the weights of the different greenhouse gases. The "curvature" of the function describing the costs of climate change is also important. By this we mean the following: Assume that a temperature increase of 3 degrees gives damages/costs corresponding to 2% of world GDP (this cost *level* is of no importance for the relative weights of different greenhouse gases). The curvature of the cost function for climate change is a parameter which tells us what the cost of a temperature increase of 6 degrees is. If the cost function for climate change is linear, a 6 degree temperature increase gives a cost which is twice as large as the cost of a 3 degree temperature increase, i.e 4% of world GDP. However, it can be argued that the cost of a 6 degree temperature increase is likely to be considerably higher than twice the cost of a 3 degree temperature increase.

We consider two alternative functions: In the one with a weak curvature, a temperature increase of 6 degrees is assumed to cost $2^{1.5}=2.8$ as much as the cost of a 3 degree temperature increase (i.e. 5.7% of GDP for the example above). The function with the strong curvature implies a cost of $2^3=8$ as much as the cost of a 3 degree temperature increase (i.e. 16% of GDP for the example above).

Intertemporally Efficient Emissions of Greenhouse Gases

At any time t, the emissions of the greenhouse gases is given by the vector $x(t) = (x_1(t),...,x_n(t))$. The climate at time t is summarized by the increase in average global temperature above its preindustrial level, denoted by $T(t)$. World income (I) is at any time assumed to depend on emissions of the different greenhouse gases $(x(t))$, of the climate $(T(t))$, and on a number of exogenous factors which we do not explicitly specify. Thus

(1) $$I(t) = I(x(t), T(t), t)$$

It is assumed that I is declining in T. For any given T and t, the emissions maximizing I are denoted by $x^0(T,t)$. It is assumed that I is increasing in x_i for $x_i < x_i^0(T,t)$, or, alternatively stated, emission abatement is not free. The partial derivatives $\partial I/\partial x_i$ represent the marginal costs of reducing emissions of the greenhouse gases.

The atmospheric concentration of greenhouse gas i is denoted by $S_i(t)$[1]. We measure atmospheric concentration in the same units as emissions per year. Transformation to more conventional measures such as "parts per million" is straightforward [2]. The greenhouse gases are assumed to develop according to the following differential equation

(2) $$\dot{S}_i(t) = x_i(t) - \delta_i S_i(t)$$

The parameter δ_i represents "natural depreciation" of greenhouse gas i. The type of "radioactive decay" assumed in (2) (i.e. δ_i constant) is a reasonable approximation for most greenhouse gases, at least as long as indirect effects are ignored. For CO_2, however, the process of removal of CO_2 from the atmosphere is more complex than suggested by (2). Nevertheless, we shall use equation (2) for all greenhouse gases in this and the next section. The numerical analysis is based on a more complex and more correct description of the change of atmospheric concentration of CO_2 and some of the other greenhouse gases. In particular, in this section it is assumed that δ_{CO_2} decreases with time as suggested by the IPCC, see Houghton et al. (1992).

The development of the global average temperature depends on the development of radiative forcing. The global average temperature is a lagged, increasing function of radiative forcing. More precisely, we assume that this relationship is of the type

(3)
$$\dot{T}(t) = \sigma\left[\lambda\Sigma_i h_i(S_i(t)) - T(t)\right]$$

where $h_i(S_i)$ is the increase in radiative forcing from greenhouse gas no. i since its preindustrial level (measured in W/m^2). The functions h_i vary between greenhouse gases, see Houghton et al. (1990). In particular, the h-function for CO_2 is of type $ALn(S)+B$, while the h-functions for CH_4 and N_2O are of the type $A(\sqrt{S})+B$ (where A and B are constant parameters which differ between the gases)[3]. All CFC-gases have h-functions of the type AS, where the constant parameters A depend on which CFC gas we have.

The parameter λ is the factor of proportionality between radiative forcing and the long-run temperature response. This factor of proportionality is uncertain. In our numerical analysis we have set $\lambda = 0.75$, i.e. we have assumed that an increase of radiative forcing of 1 W/m^2 gives a long-run temperature increase equal to 0.75 degrees (celsius). This relation is based on the "best estimate" of climate sensitivity to radiative forcing as given by the 1992-report of the IPCC (Houghton et al., 1992).

The parameter σ represents the response time for the climate system. In the numerical analysis it is assumed that $1/s=40$ (years), i.e. s=0.025. However, to simplify the exposition, the time lag between the actual temperature increase and the increase in radiative forcing is ignored in this and the next section. In other words, we have set $1/\sigma=0$, which from (3) implies that

(4)
$$T(t) = \lambda\Sigma_i h_i(S_i(t))$$

The socially optimal time paths of emissions of greenhouse gases are found by maximizing

(5)
$$V = \int_0^\infty e^{-rt} I(x(t), T(t), t)\, dt$$

subject to (2) and (4) (and given start-values for all S_i). In (5), r is an exogenous discount factor, which for simplicity is assumed constant. A further discussion of this discount factor is given in the numerical analysis section. A straightforward use of optimal control theory gives the following necessary conditions for a social optimum, where the $q_i(t)$'s are the costate variables (measured positively) of the differential equations (2), i.e. the shadow prices of the atmospheric concentrations of greenhouse gases:

$$\dot{q}_i(t) = (r + \delta_i)q_i(t) + I_T(x(t),T(t),t)\lambda h_i'(S_i(t)) \quad (6)$$

$$I_{xi}(x(t),T(t),t) = q_i(t) \quad (7)$$

where we have used the notation $I_T = \partial I/\partial T$ and $I_{xi} = \partial I/\partial x_i$. Equation (7) tells us that the marginal cost of reducing greenhouse gas i should be equal to the shadow price of this greenhouse gas. Equation (6) give the development of these shadow prices, and may be solved to give

$$(8) \quad q_i(t) = \int_t^\infty e^{-r(\tau-t)}\left[-I_T(x(\tau),T(\tau),\tau)\right]\left[e^{-\delta_i(\tau-t)}\lambda h_i'(S_i(\tau))\right]d\tau$$

The shadow price $q_i(t)$ thus has a straightforward interpretation: It is the present value of the marginal damage of the temperature increase at all future time points caused by the emission of one unit of greenhouse gas no. i. In (8), the first term in square brackets gives the damage per unit of temperature increase, while the second term in square brackets gives the temperature increase in time τ caused by one unit increased emissions at time t.

The shadow prices $q_i(t)$ are equivalent to the greenhouse gas weights discussed in the introduction: If $q_i(t)/q_{CO_2}(t) = \alpha_i$, then gas no. i should be treated in the same way as α_i tons of CO_2 in period t. This follows from (7), which states that the efficient abatement level of greenhouse gas no. i in period t should be so large that the marginal costs of further emission reductions are α_i times as large (per ton) as the marginal costs of reducing CO_2 emissions. Imagine for a moment that all δ_i's are equal and that $h_i'(S_i) = \alpha_i \cdot h_{CO_2}'(S_{CO_2}(t))$ for all t in equilibrium[4]. If these conditions held, it is clear from (8) that $q_i(t) = \alpha_i q_{CO_2}(t)$ for all t, i.e. the relative shadow prices, or weights, of greenhouse gases would be constant. However, in reality the δ_i's vary strongly among different greenhouse gases, e.g. 0.091 for CH_4, 0.007 for CO_2 and N_2O, and from 0.002 to 0.556 for the different CFC's and HCFC's. As mentioned previously, the h_i-functions are also non-linear and differ across greenhouse gases. The relative shadow prices $q_i(t)/q_{CO_2}(t)$ will therefore in reality vary over time.

The shadow prices $q_i(t)$ are derived above for the optimal emission path. However, they have the same interpretation as above for an arbitrary path of greenhouse gas emissions, with a corresponding temperature development: $q_i(t)$ gives the present value of the marginal damage of the temperature increase at all future time points caused by the emission of one additional unit of greenhouse gas no. i at time t. Or, to be more precise, consider an arbitrary reference path, and a perturbation of this path given by $x_i(t)$ being changed to $x_i(t)+\Delta x_i$ for a short time interval $[t,t+\Delta t]$. The change in V following from such a permutation can be shown to be

$$(9) \qquad \Delta V = \left[I_{xi}(x_i(t),T(t),t) - q_i(t)\right]\Delta x_i \Delta t$$

Here the term $I_{xi}\Delta x_i\Delta t$ represents the direct income gain from increasing emissions of greenhouse gas i, while $q_i\Delta x_i\Delta t$ represents the costs of the climate change associated with this increase in emissions. The optimality condition (7) simply states that along an optimal emission path, it is not possible to increase V through any permutation of the original emission path.

The Stationary Equilibrium

Consider the case in which the income function (1) is independent of time[5]. In this case there is a stationary equilibrium, which the optimal variables converge to. The stationary equilibrium is found by setting $S_i=0$ and $q_i=0$ in (2) and (6). Denoting stationary variables by an asterix, it is clear from (2)-(7) that

$$(10) \qquad q_i^* = -I_T^* \frac{h_i'(S_i^*)}{r+\delta_i}$$

$$(11) \qquad I_{xi}(x^*,T^*) = q_i^*$$

$$(12) \qquad x_i^* = \delta_i S_i^*$$

$$(13) \qquad T^* = \lambda\Sigma_i h_i(S_i^*)$$

The equations (10)-(13) consist of $3(N+1)+1$ equations determining the $3(N+1)+1$ variables q_i^*, x_i^*, S_i^* and T^* (i=0,1,...,N).

From (10) we see that

$$(14) \qquad \frac{q_i^*}{q_{CO2}^*} = \frac{h_i'(S_i^*)}{h_{CO2}'(S_{CO2}^*)} \frac{r+\delta_{CO2}}{r+\delta_i}$$

The stationary weight of any greenhouse gas relative to CO_2 thus depends on this gas's radiative forcing compared with CO_2, and also on its lifetime.

The relative weights $q_i^*/q_{CO_2}^*$ depend on physical characteristics of the greenhouse gases, namely the h_i-functions and the Δ_i-values. However, they also depend on economic variables, namely the discount

TABLE 9.1: Relative Weights of Greenhouse Gases in a Stationary Equilibrium (with r=0.005)

Greenhouse gas	Lifetime $(=1/\delta_i)$	Relative radiative forcing $(=h_i'/h_{CO_2}')$	Relative stationary weights $(=q_i^*/q_{CO_2}^*)$
CO_2	150 [6]	1	1
CH_4	11	58	7
N_2O	150	206	206
CFC-11	60	3970	2138
CFC-12	120	5750	5031
CFC-113	100	3710	2886
CFC-114	220	4710	5757
CFC-115	550	4130	7067
HCFC-22	16	5440	940
HCFC-123	1.7	2860	56
HCFC-141b	11	2900	353
HCFC-142b	22.4	4470	1051
HFC-134	15.6	4130	697
HFC-152a	1.8	4390	91
CCl_4	47	1640	728
CH_3CCl_3	7	900	71

rate r and the S_i^*-values (since the h_i-functions are non-linear). From (10)-(13) it is clear that the S_i^*-values depend on the I-function, which is an aggregate description of important characteristics of the economy.

We have not made any attempt to specify the I-function, and can therefore not calculate S_i^*-values along an optimal path. The relative weights of the different greenhouse gases along such an optimal path can therefore not be calculated. However, in order to get some feel of the sizes of $q_i^*/q_{CO_2}^*$, Table 9.1 gives these relative weights for the h_i'/h_{CO_2}'-ratios at current greenhouse gas concentrations. In this table, we have set r=0.005. With the interpretation of r as an interest rate minus an exogenous rate of income growth (see note 5), this value is consistent with what is assumed in the next section.

Numerical Analysis

The stationary equilibrium of the previous section is useful to illustrate which factors affect the relative weights of different greenhouse gases. However, for numerical estimates of these weights the stationary equilibrium is of limited interest. There are two reasons for this. In the first place, the income function will in reality depend on time (and usually in a more complex way than considered in note 5). A stationary equilibrium will therefore in general not exist. In the second place, and more importantly, even if a stationary equilibrium exists, it will take several decades to be reached. The stationary equilibrium is therefore not relevant for climate policy during the next two or three decades.

Little can be said in general outside the stationary equilibrium. We therefore turn to a numeric analysis for a time horizon of 240 years. As mentioned previously, we have not made any attempt to estimate the income function. A complete description of the social optimum is therefore impossible. What we do instead is to calculate the relative shadow prices $q_i(t)/q_{CO_2}(t)$ for an exogenously specified path of emissions of all greenhouse gases. These time paths are as follows: Emissions of CO_2, CH_4 and N_2O are assumed to develop so that the yearly growth rates of the atmospheric concentrations of these compounds are 0.4%, 0.8% and 0.3%, respectively, until 2020. This corresponds to the most recent observations of global growth rates of the compounds, see WMO (1992). After this time emissions of CO_2 are assumed to decline by 0.2% per year. CH_4 and N_2O are only partly affected by man-made emissions, and the sources are poorly known, which makes control of emissions uncertain. We therefore assume continued growth of atmospheric concentration, although at a slow rate (0.1% per year). The emissions of all other greenhouse gases (chlorine compounds) are assumed to decline towards 2020, after which emissions are zero. This should be a reasonable assumption, since the "Montreal Protocol" requires an efficient phaseout of ozone depleting substances over the next 20-30 years.

An exogenous x(t)-vector (for all t) gives a particular development of all $S_i(t)$, and therefore also a particular development of the climate (represented by T(t)). From (8) we can then calculate all $q_i(t)$'s, which we can write as

$$(15) \qquad q_i(t) = \int_t^\infty e^{-r(\tau-t)} \left[-I_T(x(\tau), T(\tau), \tau) \right] \theta_i(\tau, t) d\tau$$

where the second square bracket in (8) now is written as $\theta_i(\tau, t)$. This term, which depends on the whole time path of atmospheric concentrations and global average temperature between t and τ, tells us how much the temperature rises at τ as a consequence of 1 unit of additional emissions of greenhouse gas i at time t. We have calculated $\theta_i(t,t)$ along our reference path for the 16 greenhouse gases listed in Table 9.1.

The income function is specified as

$$(16) \qquad I(x,T,t) = \left[I - D(T) \right] Y(x) e^{\gamma t}$$

where $Y(x)e^{\gamma t}$ is potential income, i.e. income in the absence of adverse climate effects. D(T) is the damage function, with D(0)=0, D'(T)>0, and D''(T)≥0. D(T) tells us what portion of potential income is lost due to the adverse effect of climate change. This specification gives

(17) $$-I_T(x,T,t) = D'(T)Y(x)e^{\gamma t}$$

Inserting (17) into (15) gives

(18) $$q_i(t) = Y(x)e^{\gamma t}\int_t^\infty e^{-(r-\gamma)(\tau-t)}D'(T(\tau))\,\theta_i(\tau,t)\,d\tau$$

The appropriate discount rate r is determined as follows. Assume that the true intertemporal objective function is

(19) $$W = \int_0^\infty e^{-\rho t}N(t)\,u\!\left(\frac{I(t)}{N(t)}\right)dt$$

where N(t) is population, u is a utility function, and p is a utility discount rate. (In the specification above, it is implicitly assumed that consumption is proportional to income.)

Consider a change in the time path of total income, denoted by ΔI(t). The associated change in welfare follows from (19):

(20) $$\Delta W = \int_0^\infty e^{-\rho t}u'\!\left(\frac{I(t)}{N(t)}\right)\Delta I(t)\,dt$$

Denote the discount factor in front of ΔI(t) by β(t), i.e.

(21) $$\beta(t) = e^{-\rho t}u'\!\left(\frac{I(t)}{N(t)}\right)$$

Comparing (5) with (20) and (21), it is clear that the discount rate r in (5) corresponds to -β̇/β in (20) and (21). Moreover, using the approximation that I(t) grows at the rate γ, and that population grows at the rate n, it follows from (21) that

(22) $$-\frac{\dot{\beta}(t)}{\beta(t)} = \rho + \omega(t)(\gamma - n)$$

where

(23)
$$\omega(t) = -\frac{u''\left(\dfrac{I(t)}{N(t)}\right)}{u'\left(\dfrac{I(t)}{N(t)}\right)}\frac{I(t)}{N(t)}$$

Although ω in principle may vary over time as I/N changes, we shall assume (as is usually done in analyses of this type) that ω is a constant parameter.

Inserting r=-β/β and (22) into (18) gives

(24)
$$q_i(t) = Y(x)\,e^{\gamma t}\int_t^\infty e^{-R(\tau-t)}D'(T(\tau))\,\theta_i(\tau,t)\,d\tau$$

where

(25)
$$R = \rho + (\omega - 1)g - n$$

and g=γ-n is per capita income growth.

Consider first the value of the term r in (25). This term represents discounting purely because of time. From an ethical point of view, it is difficult to defend a large value of ρ. If ρ=0.03, for example (as used by e.g. Peck and Teisberg, 1992), and we assume that the time between two generations is 30 years, then each generation is given only 41% of the weight of the previous generation. Even with ρ=0.01, each generation is given only 74% of the weight of the previous generation. Equal weight to each generation implies that ρ=0. We shall use ρ=0 as our base case, but also consider ρ=0.01.

In a long-run analysis of the current type, the term w (=(I/N)u''/u') in (25) represents society's attitude towards the distribution of consumption between generations. The more weight society gives to equity, the higher the value of ω. The values used in economic analyses are often in the range 1-3. The logarithmic utility function, used by e.g. Peck and Teisberg (1992) and Nordhaus (1992), has ω=1. Scott (1989) has estimated ω to be 1.5 for the United Kingdom, which is also the value used by Cline (1993), and is the one we use in our analysis.

The population growth in the 80's was 1.8%. We shall assume that population grows by 1.25% for the first 50 years, and that the growth rate then declines to 0.25% for the following 50 years. After the year 2090

TABLE 9.2: The Value of R for $\omega=1.5$ and $\rho=0$.

	1991 – 2040	*2041 – 2090*	*2091 – 2230*
n	1.25%	0.25%	0
g	1.5 %	1.0 %	1.0 %
$R=\omega+(\omega-1)g-n$	- 0.5 %	0.25%	0.5 %
R+g+n	2.25%	1.5 %	1.5 %

population is assumed to be constant. These growth rates give a population of about 10 trillion in 2050, and about 11 trillion by 2100. These assumptions are roughly in line with other projections, see e.g. Houghton et al. (1990).

The average per capita growth of the world gross product, which we may use as a proxy for g(t), was 1.2% in the 80's. We assume that the per capita growth will be 1.5% for the next 50 years, after which it declines to 1% for the remaining 190 years our analysis covers.

Table 9.2 summarizes our assumptions about the terms affecting R (given by (4.11)). In this Table we have also given the interest rate (=R+g+n).

The function D(T) in (16) is specified as AT^a. The parameter A is of no importance for the present analysis, while the interpretation of the parameter a was given in the end of the introduction. Our main alternative is a=1.5, which is roughly in line with the estimates of Cline (1992). We also consider the case of a=3, which implies that as the global average temperature increases, additional increases cause severe damage.

Results

In Table 9.3, we have given the relative weights $q_i(0)/q_{CO_2}(0)$ for the 15 non-CO_2 greenhouse gases which follow from our formula (24). In the first column, we have repeated the lifetimes from Table 9.1. The second column gives the weights for our main case, i.e for R given by Table 9.2, and a=1.5. The third column gives weights for the case of a=3, i.e. a strongly curved damage function, while the fourth column gives weights for R-values 0.01 above the values of Table 9.2. In addition, we have repeated the stationary equilibrium values from Table 9.1, and have given the "global warming potential" (GWP) as calculated by the IPCC for a time horizon of 100 years, see Houghton et al. (1992). These GWP-figures follow from a formula similar to (24), except that (a) the horizon is 100 years, (b) R=0, (c) D'(T(t))=1, and (d) the terms $\theta_i(t,t)$ give the increase in radiative forcing at time t as a consequence of 1 unit of additional emissions at time t.

Consider first the greenhouse gases with short lifetimes. It is clear from

TABLE 9.3: Relative Weights of Greenhouse Gases for Four Alternative Sets of Economic Assumptions

Green-house gas	Lifetime $(=1/\delta_i)$	R given by Table 2 $a=1.5$	R given by Table 2 $a=3$	R 0.001 higher than in Table 2 $a=1.5$	Weights from Table 1	GWP from IPCC
CO_2	120-300[7]	1	1	1	1	
CH_4	11	13	7	19	11	11
N_2O	150	316	348	289	323	270
CFC-11	60	3593	3259	3819	3602	3400
CFC-12	120	8245	8413	8038	8383	7100
CFC-113	100	3640	3737	3531	3702	4500
CFC-114	220	7458	8791	6413	7642	7000
CFC-115	550	9401	11880	7530	9629	7000
HCFC-22	16	1709	937	2397	1476	4200
HCFC-123	1.7	40	17	62	30	90
HCFC-141b	11	440	224	640	364	580
HCFC-142b	22.4	1802	1091	2409	1627	1800
HFC-134	15.6	1079	588	1509	929	1200
HFC-152a	1.8	160	72	247	122	150
CCl_4	47	1060	852	1218	1040	1300
CH_3CCl_3	7	73	34	110	58	100

Table 9.1 that for all gases with lifetimes less than 25 years, the weights depend quite significantly on what is assumed about the curvature of the damage function and about the interest rate. A couple of these gases, namely HCFC-22 and HCFC-123, also have a considerably lower weight in our main alternative than what is suggested by the IPCC. The opposite is true for the greenhouse gas in our analysis which has the longest life-time: CFC-115 has a considerably higher weight in all of our three cases than what is suggested by the IPCC.

In spite of the observations above, the general impression one gets from Table 9.3 is that the Global Warming Potentials given by the IPCC (for a 100 year horizon) are not too different from the weights we calculate. In fact, for most gases, the IPCC-numbers are within the ranges given by our three alternative sets of assumptions. This is quite surprising, given the differences in methods for calculating the weights.

Notes

1. $S_{CO_2}(t)$ stands for the atmospheric concentration of CO_2 above its pre-industrial level.

2. The formula is as follows: 1 gigaton $(=10_{15}$ gram) emissions of gas i is equivalent to $(6.84/M_i)$ ppm (parts per million) atmospheric concentration, where M_i is the molecular weight of gas i. For instance, we have $M_{CO_2}=44$, so that 1 giga-

ton of CO_2 emissions gives an increase in the atmospheric concentration of CO_2 equal to $6.84/44 = 0.156$ ppm.

3. For CH_4 and N_2O the relationship between atmospheric concentrations and radiative forcing are not additive as assumed in (3). However, this additive form is a reasonable approximation also for these gases, see Houghton et al. (1992) for details. In the numerical calculations we use the expressions from Houghton et al. (1992).

4. This latter condition would hold if all h_i-functions were linear. For the non-linear case it would hold only by chance.

5. Alternatively, time could enter separably, i.e. $I=I(x,T) \cdot f(t)$. If $f(t)=e^{\gamma t}$ the analysis in this section remains valid, except that r everywhere must be replaced by r-γ.

6. The lifetime of CO_2 in the numerical analysis is assumed to increase with time, from 120 years to more than 300 years. This is in accordance with the suggestions by the IPCC, see Houghton et al. (1990, 1992). The constant lifetime of 150 years used in Table 9.1 is thus an approximation to the more complex relationship used in the numerical analysis.

7. See note 6.

References

Cline, W. 1992. *The Economics of Global Warming*, Washington: Institute for International Economics.

Cline, W. 1994. "Socially Efficient Abatement of Carbon Emissions", This Volume.

Houghton, J.T., Jenkins, G.J. and Ephraums, J.J., eds. 1990. *Climate Change, The IPCC Scientific Assessment*. Cambridge University Press.

Houghton, J.T., Callander, B.A. and Varney, S.K., eds. 1992. *Climate Change 1992, The Supplementary Report to the IPCC Scientific Assessment*. Cambridge University Press.

Nordhaus, W. 1992. "The 'DICE' Model: Background and Structure of a Dynamic Integrated Climate Model of the Economics of Global Warming." New Haven: Yale University, mimeo.

Peck, S.C. and Teisberg, T.J. 1992. "CETA: A Model for Carbon Emissions Trajectory Assessment, *The Energy Journal* 13, 55-77.

Scott, M.F. 1989. *A New View of Economic Growth*. Oxford: Clarendon Press.

WMO 1992. Scientific Assessment of Ozone Depletion: 1991, WMO Global Ozone Research and Monitoring Project, Report No. 25.

10

The Need for Cost-Effectiveness and Flexible Implementation of the Climate Convention and Subsequent Protocols

Ted Hanisch

Introduction

When the Convention on Climate Change was signed in Rio in June 1992 only 5 years had passed since the Brundtland Commission[1] pointed to the danger of global warming, and 22 months had passed since the first scientific basis for action was ready from the IPCC[2]. In October 1990, the Second World Climate Conference put the issue higher on the political agenda, and by December 1990, the UN General Assembly established the INC[3] with the mandate to negotiate a convention on climate change. Really, the international response to climate change has been fast. However, the question remains: Can we expect the Convention to produce adequate solutions or at least some significant steps towards protecting the atmosphere?

In this chapter we argue that the answer is a qualified yes, provided that one is familiar with the causes of climate change and the implications of response strategies. If the name of the game had been negotiating a Convention on New Energy Systems for the World, maybe expectations would have been more realistic and the assessment of the convention would have been more positive. Before and during the Rio Summit many environmentalists and experts argued simply that the climate convention was weak and vague. However, compared to other multilateral regimes of similar complexity, such as GATT or UNCTAD, the start was not at all bad. The Rio Convention entails some very promising and interesting elements and may be seen as the first in a new generation of environmental agreements. What's more, it may gradually develop into an effective regime.

A commitment from countries that would procure a halt, or perhaps a reduction in total global emissions of Greenhouse gases (GHGs), was never on the agenda for the INC. The most binding set of commitments proposed but not agreed upon was a stabilization of emissions at 1990 levels by developed countries, individually or jointly, by the year 2000. If, hypothetically speaking, this case had been agreed upon, global emissions would still have continued to grow as a consequence of increased energy (e.g.,coal) consumption by developing countries. As pointed out by David Pearce[4] a stabilization of all GHG-emissions at the present level will only reduce the 0.3°C increase in global temperature per decade projected in the IPCC´s BAS-scenario to 0.23°C. The IPCC itself calculated in its first report that a 60 percent reduction in global emissions was necessary to stabilize the atmospheric concentrations of GHGs by the middle of the next century. There is no scenario, however, in IPCC's first and second report projecting such a sharp cut in emissions.

The issue of climate change raises problems for a regulatory regime because major national interests, particularly concerning energy supply and competitive ability, may be at stake. Since there are costs connected to the effective measures that each country may take, governments will be anxious to avoid free-riding from competitors. Free-riding is only one aspect of complicated burden-sharing issues that arise when countries on different levels of development with different energy systems, try to agree on a joint effort to reduce emissions, mainly from the burning of fossil fuels. The difficult issue of equity is well known from other multilateral negotiations.

Because of the long time span between measures and their ensuing effect(s), it is also hard for governments and parliaments to convince their voters that they will have to pay now to solve a problem that will occur in a hundred years – without knowing how serious the problem is and what real effect taxes or other measures will actually have. In politics five years is a generation; the difference between eternity and half a century is negligible.

One should not expect a convention negotiated in 15 months to be more than a modest step in the construction of an effective climate policy regime. One reason expectations were so high may be due to the success, in September 1987, of negotiating the Montreal Protocol on Substances that Deplete the Ozone Layer. Environmental NGOs and negotiators to a large extent moved from ozone to climate change, many of them with the idea that the second shot would be much like the first.

As negotiations in Rio proceeded, more delegates had a chance to reflect upon the complexity of the issue and more economic studies that entailed some calculations of short term costs were made available to them. What was more important than the cost issue as such was an emerging understanding that a detailed specification in the Convention of measures to be taken by each and one country signing up was not conceivable. The reason, of course, is that the sources of climate change are to be found in most sectors of societies. And, what is equally important, these sources have a different mix in each country. A usual standardization of measures would then be a major interference in each country's sovereign right to form its policies, including energy policy.

On a very general level there was, from the beginning, a strong argument for flexibility if a country is to meet its commitments under the Convention. One of the first suggestions for bringing this into consideration was the proposal from the USA about the so-called "comprehensive approach". Originally, this was meant inter alia to allow for measures taken to reduce emissions of ozone depleting substances to also be counted as abatement measures under the climate convention.

Apart from this strategy, which failed, the proposal actually entailed an important principle, which possibly contributed to cost-effectiveness. A convention based on this approach would allow parties to give priority to those greenhouse gases in their national inventory where their overall cost-benefit analysis would find it to be the best solution. The idea was strongly supported by a number of delegations to the INC. From Norway the idea was picked up and argued that it should be enlarged to a more general approach that would introduce mechanisms for cost effectiveness, not only across sources and sinks, but also across sectors and borders[5].

These ideas were presented formally by Norway and later by Germany to the INC, with the introduction of a mechanism for "joint implementation" of commitments. In reality this was a proposal to introduce a negotiated system for leasing emission permits. Rather than allowing for trade, the proposal was to build an intergovernmental mechanism that would produce most of the same benefits, without the need for perfect market conditions.

By definition all commitments under a Climate Convention will be national. The provision for joint implementation will not in any way interfere with the issue of national responsibility. It will mainly serve to open a new arena for cooperation in abatement strategies, based on mutual benefits.

Commitments – The Proof of the Pudding

There are universal and specific commitments in the Convention. The latter refer only to developed country parties. The universal commitments relate mainly to the creation of national inventories on sources and sinks and to the securing of a flow of information and knowledge.

The main purpose is to introduce transparency concerning the causes of climate change, relevant technologies to limit emission of GHGs, and cooperation in the further development of such knowledge.

All parties also commited themselves in very general terms to protect sinks, limit emissions of GHGs, prepare for adaptation to the impacts of climate change, and cooperate where appropriate in these efforts.

The specific commitments relate to developed countries, as listed in a specific annex.[6] The crucial paragraph 2a in Article 4 is long, but its substance is a political declaration of a joint goal for developed countries to break the trend of increased emissions of CO_2 and other anthropogenic GHGs:

> Each of these Parties shall adopt national policies and take corresponding measures on the mitigation of climate change, by limiting its anthropogenic emissions of greenhouse gases and protecting and enhancing its greenhouse gas sinks and reservoirs. These policies and measures will demonstrate that developed countries are taking the lead in modifying longer term trends in anthropogenic emissions consistent with the objective of the Convention, recognizing that the return by the end of the present decade to earlier levels of anthropogenic emissions of carbon dioxide and other greenhouse gases not controlled by the Montreal Protocol would contribute to such modification....

Presumably few readers will be able to decipher the text as cited above in less than five minutes. The substance, in summary, is that here is a political commitment to break the trend of increased emissions by the end of this decade. According to Article 12, all Parties shall communicate information to the Conference of the Parties on:

1. A national inventory of anthropogenic emissions by sources and removals by sinks, and
2. A general description of steps taken or envisaged to implement the Convention.

Developed countries shall, in addition, communicate a detailed description of the policies and measures that they have adopted and a specific estimate of the effect that the polices have had. These parties are requested to make this information available within six months after the Convention's entry into force, while developing countries have a respite of three years or more.

According to Article 4, paragraph 3, developed country parties are committed to cover the full incremental costs of developing countries in providing the information mentioned; likewise for their fulfillment of general commitments.

Furthermore, the Convention introduces flexible elements that are new to international environmental regulation. The joint implementation of commitments is accepted. This means that one country whose marginal abatement costs are high, can look for another country party where the same limitations or reductions can be obtained at a lower cost. Most abatement measures will have other effects that are economically and environmentally positive. There is a basis for mutual interest. In the long run, total investments in abatement measures will be moved towards regional and global cost-effectiveness.[7]

In addition, the Convention recognizes the need for harmonization of measures; e.g. carbon taxes to avoid trade distortions. However, based on the negative experience of the International Energy Agency in its attempt to harmonize energy policies and taxes in the past, one should not expect fast progress in the short term.

The Convention also accepts the idea of a "comprehensive approach", whereby each party in the longer term is free to choose which gases or sinks should be given priority. Altogether, parties will have an increased flexibility whereby cost-effective or low-cost packages of measures can be constructed.

The Problem of Burden Sharing

Various policies and investments are available to governments and industries to decrease greenhouse gas emissions per unit of energy used. A number of macro-economic studies indicate that, for most developed OECD-countries, a stabilization could be achieved at no cost or in many cases prove to be economically profitable.[8] However, one should note that these calculations assume implementation of rather tough carbon tax policies and even tougher structural adjustment policies than experienced in recent years. Existing subsidies to energy production of the type we find in a number of countries will have to be removed first. Severe limitations for developing countries would cause very substantial economic setbacks, given the energy and energy-related technology available today or in the near future.

One cannot scientifically assess what is or is not politically feasible. It may be noted, however, that very serious political conflicts have arisen from less serious economic risks than the existing implications of a new climate regime. Some governments were anxious to avoid political disturbances on top of economic costs. A favorite example among European negotiators is the fact that it would make sense for the USA to put a carbon tax on oil and gasoline to support energy efficiency and limit imports of fossil fuels. The public reaction to such a proposal would, however, be very negative – to say the least.

When delegations to the INC started to negotiate the Convention, many were already concerned about how a convention would impact their economies. Many delegations did not know. As the process went on, more and more governments had studies available to them on possible economic implications of limiting emissions. Discussions were slow and delegates were anxious to not put their own country in a loser's position.

A Few Practical Examples

If you negotiate an agreement on stabilization of emissions for all developed countries from year zero, one might assume this would create a fair and acceptable burden-sharing. However, once into the practical steps of limiting emission of CO_2, one will immediately see large differences exist, given the starting point from a given and common year. One country may rely to a large extent on coal for electricity production and at the same time have a tradition of low energy prices. Another country may have changed some time in the past a fair amount of its primary energy supply to natural gas and used the price mechanism to increase energy efficiency. The cost of limitation will be much higher for the latter country. Roughly speaking, one will find that countries with the dirtiest energy systems in year zero will have low or negative reduction costs, compared with those with modern and clean systems. Such differences in marginal costs vary substantially. For example, a study by DRI[9] concluded that a stabilization by 2000 would need a carbon tax of 120 USD per ton carbon for the US and 400 USD for Japan.[10]

The usual way of defining the baseline for actions taken by each party and group of parties in an international convention is the so-called grandfathering principle. It is widely used probably because it is so simple and at the same time so accepting of the status quo. In the case of climate change, however, the principle will create difficulties because marginal abatement costs will vary among countries because of differences in energy systems.

The "polluter pays principle" is to be found somewhere at the bottom of the whole effort, even if is somewhat watered down. When this principle is combined with the grandfathering principle, one might say that a new principle will appear: The dirty energy principle. According to this principle those parties who have done the least with their energy systems in the past or have based their energy supply on high polluting energy sources are to meet the lower cost obligations and vice versa.

These issues were never openly discussed during the proceedings of the INC. However, in the text itself there are traces of recognition of the underlying burden sharing issue. In the crucial paragraph 2A in Article

4 there is a direct reference to differences in marginal abatement costs when it is stated that the implementation of commitments should be done by:

> taking into account the differences in these Parties' starting points and approaches, economic structures and resource bases, the need to maintain strong and sustainable economic growth, available technologies and other individual circumstances, as well as the need for equitable and appropriate contributions by each of these Parties to the global effort regarding that objective.

So far the Convention does not give any criteria for defining what sort of differences in resource bases and which starting point would be relevant for defining specific commitments for each party and group of parties. These complicated issues are left to be negotiated at some later stage.

One might expect that subsequent protocols of the Convention will entail agreements on this. One possibility would be that the OECD group negotiated a common abatement target in such a protocol, considering differentiated obligations for parties.

Obviously, negotiating such differences is very complicated. There are good reasons why the grandfathering principle has been so widely used in the past. On the other hand, the idea of a large number of countries joining protocols of the convention that would leave them with marginal and total abatement costs two to five times that of their competitors may also seem quite naive. There is no way that a convention, which in reality is restructuring the energy systems of the world, could work on the basis of burden sharing principles that totally disregard the most obvious findings from studies of energy policy options.

Notes

1. Gro Harlem Brundtland et al. 1987. *Our Common Future*. Oxford: Oxford University Press.

2. Intergovernmental Panel on Climate Change, an expert panel established by WMO and UNEP, headed by Professor Bert Bolin. The panel was able to agree on rather complicated issues in its first report: J. T. Houghton, G. J. Jenkins and J. J. Ephramus (eds), *Climate Change, The IPCC Scientific Assessment* (Cambridge: Cambridge University Press, 1990). The findings were confirmed in J. T. Houghton, B.A. Callander and S. K. Varney (eds). 1992. *Climate Change 1992, The Supplementary Report to the IPCC Scientific Assessment*. Cambridge: Cambridge University Press.

3. Intergovernmental Negotiating committee for a Framework Convention on Climate Change, headed by Jean Ripert of France.

4. David Pearce, "Internalising Long Term Environmental Costs" p. 21, in Ted Hanisch (Ed). 1991. *A Comprehensive Approach to Climate Change*. Oslo: CICERO.

5. This was the main issue on the first CICERO seminar in June 1991 and the ideas were presented to the INC in the report from the seminar: Ted Hanisch (ed) *A Comprehensive Approach to Climate Change – Additional Elements From an Interdisciplinary Perspective*, CICERO report 1991:1.

6. Note that "developed countries" include OECD and former COMECON members. This means that no Asian country, with the exception of Japan, and no middle east country is included.

7. For a closer description of this mechanism, see T. Hanisch, R. K. Pachauri, D. Scmitt and P. Vellinga: *The Climate Convention: Criteria and Guidelines for Joint Implementation*. Oslo: CICERO Policy Note 1992:2, University of Oslo, 1992.

8. William Cline in an overview of these studies in *The Economics of Global Warming*, Institute of International Economics, Washington 1992. He concludes that cost may be zero for a 20 percent reduction in emissions if implemented over three decades. However, if one introduces reductions in emissions in the shorter run, say in the course of a decade, the costs are likely to increase .

9. *Economic Effects of using Carbon Taxes to Reduce Carbon Dioxide Emissions in Major OECD Countries*. 1992. Washington: McGraw-Hill.

10. The study by McGraw-Hill was certainly not designed to underestimate the possible burden on the US economy, so one would expect the difference between US and Japan to be at least as big.

11

How Important Are Carbon Cycle Uncertainties?

T. M. L. Wigley

Abstract

This chapter summarizes the factors that determine future CO_2 concentration changes, quantifies uncertainties in concentration projections, and assesses the importance of these uncertainties in predicting future climate change. The 1992 IPCC emissions scenarios are used to determine future changes in CO_2 concentration, radiative forcing and global-mean temperature, and uncertainties accruing at each calculation stage are estimated. The main factor determining temperature change uncertainty is the climate sensitivity (ΔT_{2x}); uncertainties due to this factor are some 3-5 times larger than uncertainties arising from carbon cycle modelling uncertainties. Uncertainties arising from ΔT_{2x} are always larger than those due to uncertainties in emissions. This is particularly so over the next 50 years, during which time emissions-related uncertainties in global-mean temperature change are small. By 2100, ΔT_{2x} and emissions-related uncertainties become comparable. The relative importance of carbon cycle and emissions-related uncertainties also depends on time, with the former being dominant to about 2040. Carbon cycle uncertainties are largely controlled by uncertainties in the magnitude of the so-called "missing sink". In terms of research priorities, it is not the associated uncertainties that should be the determining factor, but the potential for (and cost-effectiveness of) reducing these uncertainties. The greatest scope for uncertainty reduction, at least in the short term, appears to lie in the area of carbon cycle modelling.

Introduction

Because of the cocktail of trace gases that mankind is emitting into the atmosphere in ever-increasing amounts, major changes in climate are expected over coming decades and centuries. The main gases concerned are carbon dioxide (CO_2), methane (CH_4), nitrous oxide (N_2O) and a large number of gases collectively referred to as halocarbons. These are all greenhouse gases; so-called because they modify the radiative balance of the Earth-atmosphere system in the way a greenhouse is (erroneously) supposed to work; i.e., by trapping some of the outgoing long-wave radiation and leading to a warming of the lower atmosphere.

Another set of gases have indirect greenhouse effects through their ability to change the oxidizing capacity of the atmosphere, and so affect the lifetimes of some of the direct greenhouse gases. This set includes carbon monoxide (CO), various nitrogen oxides (NO_x), and a number of non-methane hydrocarbons (NMHCs).

Finally, sulphur dioxide (SO_2) may have a climatic effect. When SO_2 is oxidized in the atmosphere it forms small droplets of sulphate (mainly sulphuric acid or ammonium sulphate) called aerosols. These reduce the amount of incoming shortwave radiation and have a cooling effect. Aerosols operate in two ways. In clear-sky conditions, they scatter the incoming radiation in all directions. Since a fraction is scattered back into space, the total amount of incoming radiation is reduced. Sulphate aerosols also act as cloud condensation nuclei (CCNs), causing marine clouds (which are normally CCN deficient), to have more, but smaller droplets. This, in turn, makes these clouds reflect more solar radiation back into space. The former direct aerosol effect is thought to dominate over the latter indirect effect. Aerosols also affect outgoing longwave radiation, but this effect is much smaller than their shortwave effect.

Of all of these processes, the enhanced greenhouse effect of increasing CO_2 concentrations is by far the most important, both in terms of what has happened to date and what may happen in the future. The purpose of this paper, therefore, is to summarize current understanding of the factors that determine how atmospheric CO_2 concentrations change in response to emissions of CO_2, to quantify uncertainties, and to assess the implications of these uncertainties relative to other uncertainties surrounding future climatic change. Before doing so, however, I will quantify the relative importance of CO_2 as a climate forcing agent.

Relative Importance of CO_2 in Climate Change

Changes in climate occur naturally on all time scales. This is the background upon which man-made effects are superimposed. Over the period for which instrumental data have been available, some striking changes have been observed, such as the prolonged drought in the sub-Saharan (Sahel) region of Africa. At the global scale, there has been an overall warming trend of about 0.5°C over the past century (Folland et al., 1990,

1992). We are confident that human activities have caused a substantial warming, but it is not yet possible to separate the natural and man-made components of the observed changes quantitatively (Wigley and Barnett, 1990).

Although we cannot reliably quantify the global warming (or any other aspect of past climate change) associated with different greenhouse gases, we *can* quantify their effects relatively. This is because the quantification process can be broken down into different stages. The first stage is to relate the emissions of a gas to its concentration changes. This is an area of considerable uncertainty – indeed, for CO_2, it is this stage, and the uncertainties involved, that provide the main focus of this paper. In analysing and interpreting the past, however, we can sidestep this stage because we have observational data for concentration changes.

Given the concentration data (either observed data, or values estimated from emissions) the next stage is to determine the radiative forcing changes (ΔQ) for given concentration changes (ΔC). ΔQ-ΔC relationships are known to about $\pm 10\%$ (Shine et al., 1990). At the global-mean level, knowledge of ΔQ alone is sufficient to determine the relative importance of different gases in perturbing the climate system. At smaller spatial scales, intercomparison becomes more difficult because different gases have different spatial forcing signatures and, hence, different climate response patterns (Wang et al., 1991; Kiehl and Briegleb, 1993).

We can, as a reasonable approximation therefore, assume that the relative global-mean warming amounts attributable to different gases are proportional to their global radiative forcing effects. The biggest uncertainties in estimating the climate consequences come in the next stage, linking ΔQ to climate change. Here, the uncertainty at the global scale is about a factor of three between lowest and highest estimates and considerably more at the regional scale.

Figure 11.1 shows the global-mean, top of the troposphere ΔQ values (in W/m²) for the main greenhouse gases and for sulphate aerosols over the period 1765-1990. To put these numbers in perspective, a ΔQ of 2.4W/m² corresponds to a change in the Sun's output of 1%. Such a change would be much larger than any observed or estimated solar changes over at least the past 10,000 years (see Wigley and Kelly, 1990; Hoyt et al., 1992; Kelly and Wigley, 1992).

For two of the components in Fig. 11.1, halocarbons and aerosols, a range of values is given. In the former, this is due to uncertainties regarding the radiative forcing consequences of stratospheric ozone depletion arising from halocarbon emissions (Ramaswamy et al., 1992). The two curves given in Fig. 11.1 correspond to either a zero stratospheric ozone effect (i.e., following the 1990 IPCC procedure; Shine et al., 1990) or an ozone effect equal to that calculated by Ramaswamy et al. (1992),

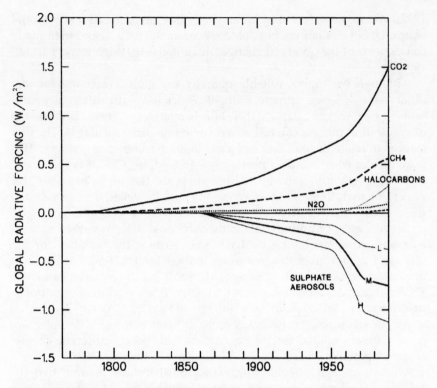

FIGURE 11.1: Global-mean radiative forcing since pre-industrial times (1765) arising from man-made emissions of CO_2 (upper full line), CH_4 (long dashes), N_2O (dots), halocarbons (thin dots or short dashes) and sulphate aerosols (full lines below zero). For halocarbons, the upper curve corresponds to the IPCC estimate given by Shine et al. (1990), while the lower estimate accounts for the negative forcing effect of stratospheric ozone depletion following Ramaswamy et al. (1992). The aerosol forcing values shown are low, mid and high estimates from Wigley and Raper (1992).

but generalized to include all halocarbons (see Wigley and Raper, 1992). The second area of uncertainty is that surrounding the aerosol effect. Here, the main difficulty is in quantifying the link between SO_2 emissions and the consequent radiative forcing changes, which leads to an uncertainty of at least ±50%. For a fuller discussion, see Wigley and Raper (1992).

The key conclusion to be drawn from Fig. 11.1 is that CO_2 is the most important contributor to man-made radiative forcing changes. Indeed, if we take the best guess values for halocarbons (i.e., including the effect of stratospheric ozone depletion) and aerosols (i.e., the central value), then the aerosol effect almost exactly cancels out the non-CO_2 greenhouse

TABLE 11.1: Past Global-Mean Radiative Forcing Changes Based On Observed Concentration Data (Updated From Shine et al., 1990, Table 2.6).

Interval	CO_2	CH_4[1]	N_2O	Halocarbons[2]	Aerosols[3]	TOTAL[4]
1765-1900	0.37	0.13	0.027	0.0	-0.12	0.41(0.53)
1765-1960	0.79	0.29	0.045	0.003(0.023)	-0.42	0.71(1.15)
1765-1970	0.96	0.38	0.054	0.006(0.083)	-0.64	0.75(1.47)
1765-1980	1.20	0.49	0.068	0.011(0.170)	-0.72	1.05(1.92)
1765-1990	1.50	0.56	0.095	0.018(0.277)	-0.75	1.42(2.43)

[1] Includes stratospheric water vapour, following Shine et al. (1990).

[2] Bracketed values ignore stratospheric ozone depletion feedback (i.e., follow Shine et al. (1990)). Other values include stratospheric ozone depletion effects calculated using the method of Wigley and Raper (1992).

[3] Best guess estimates following Wigley and Raper (1992).

[4] Bracketed values ignore both stratospheric ozone depletion and sulphate aerosols (i.e., follow Shine et al. (1990)).

effect (see Table 11.1). This, of course, is a rather oversimplified view, since the aerosol effect is largely confined to the Northern Hemisphere, and the degree of "cancelling" is strongly dependent on which hemisphere and which particular region is being considered. (Note that regional- or even hemispheric-scale cancelling of radiative forcing does not mean a cancelling of the effect on climate, since the patterns of climate response differ markedly from the pattern of radiative forcing.)

In terms of past and likely future changes, CO_2 is the most important greenhouse gas. Thus, although it may be better to base a judgement of importance on the potential for reduction in forcing rather than on the amount of forcing, from more than one point of view, CO_2 is the focal point of the greenhouse problem, just as it is the focal point of the present paper.

Description of the Carbon Cycle

The carbon cycle comprises the active reservoirs of carbon-containing material and the fluxes between them. The main net fluxes are shown in a very simplified way in Fig. 11.2. The units for amounts and fluxes are GtC and GtC/yr, where GtC means gigatonnes of carbon, and 1 Gt is 10^{15}gm. A carbon cycle model is a mathematical representation of the carbon cycle.

The biggest carbon reservoir is the deep ocean. Fortunately, this is (and is likely to remain) a net sink rather than a source of CO_2. The upper, mixed-layer of the ocean (with a global, annual-mean depth of 50-100m., highly variable both spatially and seasonally) contains about the same amount of carbon as the atmosphere, currently around 750

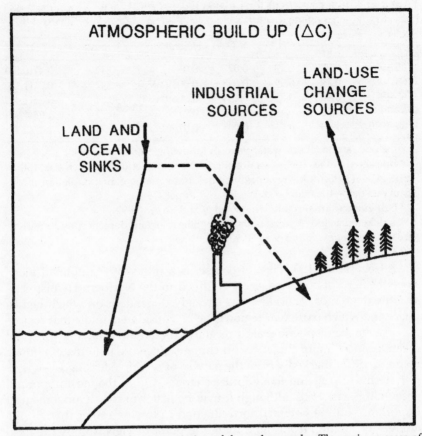

FIGURE 11.2: Simplified representation of the carbon cycle. The main sources of CO_2 are from fossil fuel combustion (I) and net land-use changes (D). The main sink is into the ocean (F). The difference between I+D-F and the rate of atmospheric increase indicates that there is a missing sink, probably into the terrestrial biosphere (dashed line).

GtC. Fluxes between ocean and atmosphere are large, with roughly 100 GtC/yr going into the ocean at high latitudes, and 100 GtC/yr coming out of the ocean in lower latitudes. In pre-industrial times, these two large fluxes must have been effectively in balance. Today, however, because of the excess CO_2 we have put into the atmosphere, this balance has been disturbed and there is currently a net flux into the ocean of 2.0 ± 1.0 GtC/yr (Watson et al., 1990; Sarmiento and Sundquist, 1992; this is the estimated mean flux over the 1980s).

There are also large fluxes into and out of the terrestrial biosphere as a result of the processes of photosynthesis, respiration and oxidation of dead plant material. A near-balance must have existed in pre-industrial

TABLE 11.2: Components of the Carbon Dioxide Mass Balance. Except Where Indicated, Figures Come From IPCC (Watson et al., 1990). Uncertainty Limits Are From IPCC.

Component	1980s values (GtC/yr)	1850-1989[1] (GtC)	Totals
2.123dC/dt or 2.123ΔC	3.5 ± 0.2[2]	140	± 10
Ocean flux (F)	2.0 ± 1.0	110	± 553[3]
Industrial sources (I)	5.4 ± 0.5[4]	213	± 204[4]
Land-use changes (D)	1.6 ± 1.0	123	± 405[5]
Budget imbalance[6]	-1.5 ± 2.7	-86	± 125

[1] i.e., 1850-1989 inclusive (140 years).

[2] Average over 1980-89 inclusive (IPCC gives 3.4±0.2 GtC/yr).

[3] As estimated using the present model. Bracketed value corresponds to a 1980s flux of 2.0 GtC/yr, with the other two values corresponding to 1.0 and 3.0 GtC/yr.

[4] From data of Keeling (1991) and Marland and Boden (1991).

[5] Based on 1850-1986 estimate of 117±35 GtC given by Watson et al. (1990, p. 13).

[6] i.e., 2.123ΔC + F - I - D : assumes that these are the only components of the balance. The negative of this number gives the magnitude of the "missing sink".

times, but this too has been disturbed by human activity. Land use changes, including deforestation and re-afforestation, have lead to a net flux from the terrestrial biosphere into the atmosphere. Currently (i.e., averaged over the 1980s), this is thought to lie in the range 1.6±1.0 GtC/yr (Watson et al., 1990). The other main perturbation of the system is that due to fossil fuel combustion and other industrial activity (mainly cement production). Over the 1980s, the mean input from this source was 5.4±0.5 GtC/yr (Marland and Boden, 1991).

When we compare the above fluxes with the current (1980s mean) rate of increase of mass of CO_2 in the atmosphere (viz. 3.5±0.2 GtC/yr), there is an apparent imbalance. Taking the central estimates, the sum of the sources (land-use changes and industrial activity) is 7.0 GtC/yr. The sum of the ocean sink and rate of atmospheric increase is, however, only 5.5 GtC/yr. This imbalance is usually referred to as the "missing carbon problem", indicating that there must be an unidentified (missing) carbon sink. Note, however, that the range of uncertainty in the apparent imbalance is large, 1.5±2.7 GtC/yr over the 1980s (see Table 11.2), a range that includes zero.

The same sort of balance calculation can be performed using estimates of the total fluxes since the middle of last century. The uncertainties are large, but the central estimates confirm the conclusion that there is probably a missing carbon sink. Table 11.2 summarizes these values.

For the land-use-change contribution, the values are based on IPCC (Watson et al., 1990). It is possible that the IPCC estimate is too low, and even more likely that the ±40 GtC error bounds are too optimistic. Data from Houghton et al. (1991) give a total land-use contribution from 1850-1989 of 138 GtC. No matter which estimates are used, however, there appears to be around 100 GtC of missing carbon.

A number of suggestions have been put forward to explain the missing carbon problem. It is likely that most, if not all of these possibilities has contributed, but their relative importance is very uncertain. The first possibility, one that I will explore further below, is that the missing carbon has been sequestered in the terrestrial biomass through the CO_2 fertilization effect.

This is the process that causes plants to grow better at higher CO_2 levels, a property that has been exploited by vegetable producers for many decades (e.g., in growing tomatoes in greenhouses with raised CO_2 concentration levels). Small-scale experiments have shown conclusively that most plants grow bigger and faster as CO_2 levels are raised. Furthermore, plant water-use efficiency increases as CO_2 increases. This effect could add to the fertilization effect by reducing the amount of time during which plants have reduced growth due to water stress. What is not known, however, is whether the results from these many small-scale experiments are applicable to natural ecosystems where plant growth is subject to a wide range of limiting factors (i.e., not just CO_2 availability).

Proving and quantifying the CO_2 fertilization effect at the ecosystem level is a daunting task. Experimentally, it would require a very large-scale study (using a CO_2-controlled environment covering a hectare or more) carried out for a long period of time (many years). Work of this kind is in progress. However, even with the biggest feasible experiment, extrapolation of results to the global scale would be equivocal. In spite of the evidence for widespread enhancement of tree growth since around 1850 (Innes, 1991), field studies (for instance, using tree-ring data) are unlikely to be entirely convincing because of alternative ways to interpret the results (Wigley et al., 1984; Innes, 1991; Briffa, 1992), and because the missing carbon need not be stored above ground. For example, a slight increase in root turnover rate would lead to a build up of soil organic matter that would be virtually impossible to detect.

There are two other missing carbon explanations that are closely related to the CO_2 fertilization effect in that they both require the carbon to be stored somewhere in the terrestrial biomass. The first is the possibility that plant growth has been stimulated by nitrogen fertilization (itself a result of industrial pollution). The second is that plant growth has been stimulated by the climate changes that have occurred over the past century or so associated with the observed global warming, with this stimulation exceeding any opposing climate effects (such as increased rate of decay of dead plant material due to warming). In both cases, the extra carbon sequestered could either be in the above ground (living) plant mass, or in the soil zone (or both); in either case, just as for the CO_2 fertilization effect, it would be virtually impossible to quantify globally through field studies.

Nitrogen fertilization is a possibility in regions where soils are nitrogen deficient (see, e.g., Innes, 1991). There is both chemical (Zöttl et al., 1989) and tree-ring (Briffa, 1992) evidence to support this, at least for Europe. In Briffa (1992), for example, a number of regions of apparently anomalous growth increases were identified, with a spatial distribution consistent with nitrogen fertilization as a contributing explanation. However, in attempting to use tree rings to identify *any* anthropogenically-enhanced physiological process, one must first factor out the natural effects of climate and plant aging, something that is exceedingly difficult to do (Briffa, 1992). Furthermore, the issue of nitrogen fertilization is complicated by interactions with the effect of CO_2. Based on small-scale experimental results, it is possible that CO_2 increases might reduce the sensitivity of trees to soil nitrogen deficiencies thereby enhancing any nitrogen fertilization effect. The primary cause in this case, however, would be CO_2.

Another piece of circumstantial evidence favouring nitrogen fertilization as a mechanism to explain the missing carbon, at least in part, is the study of Tans et al. (1990). This showed that the missing sink appeared to be largely located in the mid-to-high latitudes of the Northern Hemisphere, i.e., the zone where nitrogen oxide pollution is largest. More recent work, however, indicates that the sink is not confined to these latitudes, but extends to tropical latitudes as well (Enting and Mansbridge, 1991). This is more consistent with CO_2 fertilization or climate as the primary sink producing process.

Of all the possible explanations, climate appears to be, at best, a minor player. This is partly because even the sign of the net climate effect is uncertain (Harvey, 1989a), and partly because of the relatively small climate changes that have occurred to date. The relative stability of pre-industrial CO_2 levels also argues against a large climate effect. If climate changes since pre-industrial times had caused the terrestrial biomass to change by enough to explain the missing sink (i.e., by around 100 GtC), then pre-industrial CO_2 levels would have to have shown much larger changes than observed. (If only 50% of a 100 GtC biomass change came from the atmospheric CO_2 pool, then the atmospheric concentration would have to change by 23ppmv.)

While the weight of evidence favours CO_2 fertilization as the main explanation for the missing carbon sink, it is possible that part of the problem is accounted for by an underestimation of the amount of re-afforestation that has occurred this century (see, e.g., Kauppi et al., 1992). Estimates of the amounts of deforestation are highly uncertain, and equally great uncertainties apply to the positive side of this coin. This explanation is, however, equivalent to accepting the lower limits for land-use change contributions to the carbon budget – and these lower limits are still insufficient by themselves to restore the apparent imbalance.

Predicting Future CO_2 Levels

What is the significance of the missing carbon problem for future projections of CO_2 concentration and climate changes? To answer this question we need to consider the carbon budget in more detail. We can write this in the following way:

$$2.123 \, dC/dt = I + D - F - X \qquad \ldots(1)$$

where the left hand term gives the rate of change of atmospheric mass (the factor 2.123 converts concentration, C, in ppmv to mass in GtC), I is the industrial source, D is the net land-use source (the letter D indicates that this is largely related to deforestation), F is the flux from the atmosphere into the ocean, and X is a term representing all other factors (i.e., the missing sink term). In making future projections, I and D will be given, while F must be determined by a model of the oceanic part of the carbon cycle. The rate of increase in concentration, dC/dt, is sensitive to all of these terms and will be uncertain because of uncertainties in F (see Table 11.2). More importantly, additional uncertainties will accrue because of uncertainties in X. Future projections depend critically on how the missing sink is explained, because this determines the functional form of X(t).

One approach to solving equ. (1) is to assume that X is a constant, with a value chosen so as to balance the carbon budget today. This could be done, for instance, using best guess values of dC/dt, I and D (Table 11.2), together with a value of F appropriate to whatever ocean model is being used, to calculate the required X value.

It is clear, however, that X has varied in the past: its best guess 1980s mean value is the apparent imbalance given in Table 11.2 (1.5 GtC/yr); while the best-guess average value over 1850-1989 is only 86/140 = 0.6 GtC/yr. It is probable, therefore, that X will change in the future. An alternative, and preferred approach then is to allow X to vary by relating it to other time-varying factors, such as nitrogen emissions, climate and/or CO_2 concentration.

In the calculations described below, it is assumed that X is explained solely by the CO_2 fertilization process. In this case, the X term represents a negative feedback on the system. Concentration projections made in this way might be considered as leading to lower bound values, since future values of X under the CO_2 fertilization assumption are probably greater than would occur if other processes were significant contributors to the missing sink. However, to assert that the results are lower bound values oversimplifies the problem because, even when X is explained by CO_2 fertilization alone, one can obtain a wide range of future concentration projections for any given emissions scenario.

The way this range arises is as follows. Even when X is well-defined mathematically (which requires setting up quite a complex representation of the terrestrial part of the carbon cycle), it has a free parameter (the "beta factor", ß) which specifies how strongly plant growth increases per unit CO_2 concentration change. Beta factor values are known from small-scale experiments, but, as noted earlier, the extent of CO_2 fertilization and, hence, the value of ß at the ecosystem level is unknown. The lack of knowledge of ß means that we can balance the carbon budget even given the large uncertainties in the land use and ocean flux terms (D and F in equ. (1)), simply by changing b. For example, if D is large or F is small, then X must be large and a large value of ß (i.e., a strong CO_2 fertilization effect) is required. Uncertainties in D and F lead to a range of possible ß values, the smallest of which is close to zero, and the largest of which is close to the ß value gleaned from small-scale experiments.

A carbon cycle model designed in this way has been described by Wigley (1991, 1993). In this model, the efficiency of ocean mixing may be changed, so the strength of the ocean sink is free to be set to any chosen value. For example, the mean ocean flux over the 1980s, F(1980s), may be set at a value anywhere between the limits of 1 and 3 GtC/yr given by IPCC (Watson et al., 1990; see Table 11.2 above). Future projections require values of I and D to be specified from 1990 onwards. The model takes the 1990 values of these terms, together with the specified ocean flux, and is then run in inverse mode to calculate the value of ß required to give a balanced contemporary budget. This value of ß is then used for future projections.

Future concentration projections in this model are sensitive to the parameters controlling the strength of the two sinks, that into the ocean and that due to CO_2 fertilization, and on other assumptions made regarding the terrestrial part of the model. More generally, projections are sensitive to the magnitude of the missing sink and how this changes with time. Using CO_2 fertilization to characterize the missing sink determines the form of these future changes, but not the current magnitude of the missing sink. The procedure of "tuning" ß using both the ocean sink strength and the current deforestation source, D(1980s), has the effect that future projections for any given D(1980s) value are virtually independent of the assumed ocean sink strength (as specified, for example, by F(1980s)). If a large ocean flux is assumed, for instance, then the required value of ß will be small. The biospheric uptake of CO_2 due to the fertilization effect will therefore be small, offsetting the effect of larger ocean flux and reducing the overall sensitivity to F(1980s), a point noted by Harvey (1989b). This does not mean that projections are insensitive to the ocean flux for any fixed value of ß – far from it.

To get a range of future projections, different values of D(1980s) are used, corresponding to the range specified by IPCC for the average value of D over the 1980s. Thus, using D(1980s) = 0.6 GtC/yr leads to a small value of ß (because the apparent imbalance is small) and relatively high projected CO_2 concentrations. Using D(1980s) = 2.6 GtC/yr requires a large ß value, and the CO_2 projections are much smaller. We take these as upper and lower bound values. It is therefore the uncertainties in the strength of the missing sink (determined here by b) that lead to uncertainties in future CO_2 concentrations. Missing sink uncertainties are, in turn, determined by uncertainties in the other components of the carbon budget, particularly the ocean sink and the deforestation source strengths. An alternative upper bound could be obtained by ignoring the missing sink problem entirely. This approach is equivalent to putting ß = 0 in the present analysis, since this automatically sets the missing sink term, X, to zero. I refer to these as the no-feedback (NFB) results. This is the approach followed by IPCC in 1990, leading to the projections given in the Working Group 1 report (Houghton et al., 1990). (In fact, these projections were the average of the results from six different models, five of which were run in this mode.) If one believes that the missing sink is real, then this method must lead to an overestimate of future concentration increases.

I now give CO_2 concentration projections for four of the six CO_2 emissions scenarios given by IPCC in 1992 (Leggett et al., 1992), viz. IS92a, c, d and e. Figure 11.3 shows the emissions scenarios. Scenario ß is not considered since it is very similar to a, while f is not considered because it is bracketed by a and e. These emissions scenarios all represent possible futures under a "business as usual" or "existing policies" assumption. They differ because they assume different scenarios for future population growth, GNP growth, energy use per capita, fossil fuel resource availability, etc. While they do not represent the full range of possibilities, it is highly unlikely that future emissions will lie significantly outside their range.

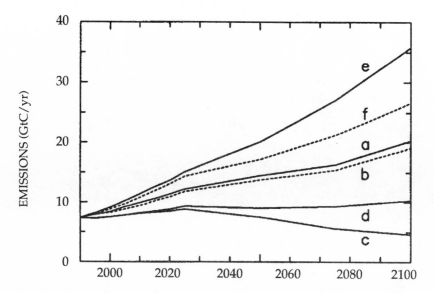

FIGURE 11.3: Scenarios for CO_2 emissions under an "existing policies" assumption, from IPCC (Leggett et al., 1992).

Figure 11.4 shows best-guess, high and low concentration projections for each scenario, together with the "no-feedback" (NFB) projection corresponding to the methodology adopted by IPCC in 1990. These projections are the same as those given in Wigley and Raper (1992). In this paper, it was assumed that the *gross* land-use flux over the 1980s was in the range 0.6–2.6 GtC/yr. A better assumption would be to use this as the range of *net* land-use emissions (where the gross to net difference is due to regrowth following deforestation): this leads to somewhat lower concentration projections than those given here. The best-guess values shown in Fig. 11.4 are not centrally placed between the high and low cases, because they use a 1990 gross land-use flux of 1.3 GtC/yr, cf Leggett et al. (1992), which corresponds to a gross D(1980s) value of around 1.1 GtC/yr, substantially below 1.6 GtC/yr and noticeably skewed towards the lower bound value of 0.6 GtC/yr.

Just how important are the uncertainties illustrated in Fig. 11.4? The concentration differences between the low and high estimates are extremely large, even larger if one takes the NFB projections as the upper bounds. For example, for scenario IS92a (which IPCC judges to be the most likely case), the high value for the change between 1990 and 2100 (viz. 410ppmv) is almost 40% above the low value of 295ppmv. This difference, however, is less alarming if one considers its consequences for global-mean temperature change.

To demonstrate this, I have taken, for each scenario, the total radia-

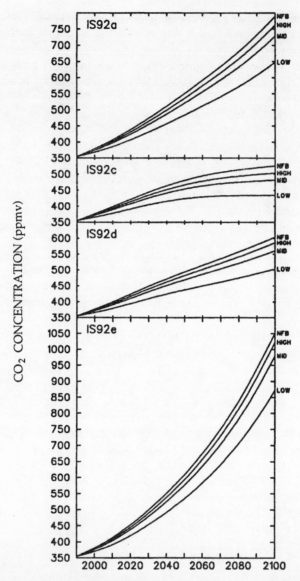

FIGURE 11.4: CO_2 concentration projections for IPCC scenarios IS92a, c, d and e. Low, mid and high values correspond to high, mid and low values of the CO_2 fertilization effect (i.e., of b) and relate to different assumptions regarding the deforestation contribution to date (1980s-mean gross deforestation fluxes of 2.6, 1.1 and 0.6 GtC/yr, respectively). The highest concentration projections shown here (NFB) correspond to the procedure adopted by IPCC in 1990 in ignoring the missing sink: obtained by setting ß = 0 and using a value for the ocean flux that reproduces the 1990 IPCC CO_2 projections (viz. a 1980s mean of 1.95 GtC/yr).

tive forcing for all other greenhouse gases and for aerosols from Wigley and Raper (1992), as updated in Wigley (1993), and added to this forcing the contribution from either the high or the low CO_2 concentration projection. Next, to convert to global-mean temperature, I have used the Wigley and Raper climate model with best-guess model parameter values (the most important of which is the climate sensitivity, specified here by the equilibrium CO_2-doubling temperature change, ΔT_{2x}). High and low temperature projections for each of the four scenarios are shown in Fig. 11.5 and results in the year 2100 are summarized in Table 11.3. (For comparison, effects on sea level rise predictions are shown in Table 11.4.) In the year 2100 (when the effect of carbon cycle uncertainties is maximized), the temperature uncertainty range for $\Delta T_{2x} = 2.5°C$ lies between 0.42°C (scenario c) and 0.46°C (scenario e). The uncertainty is almost independent of scenario because the radiative forcing uncertainties are virtually the same for each scenario (see below).

TABLE 11.3: Global-Mean Temperature Changes (°C) Over 1990-2100 For Different Emissions Scenarios (IS92a, c, d and f), Different Values of the Climate Sensitivity (ΔT_{2x} = 1.5, 2.5 and 4.5°C) and using Both Lower and Upper Bound Estimates for the Projected CO_2 Concentration Changes. Radiative Forcing Changes Due to Gases Other than CO_2 are Best Guess Values from Wigley and Raper (1992).

| ΔT_{2x} | CO_2 projection | IPCC Emissions Scenario | | | |
		IS92a	IS92c	IS92d	IS92e
1.5°C	lower bound	1.33	0.70	0.89	1.61
1.5°C	upper bound	1.64	0.97	1.17	1.91
2.5°C	lower bound	2.12	1.21	1.48	2.50
2.5°C	upper bound	2.57	1.63	1.91	2.96
4.5°C	lower bound	3.24	1.96	2.34	3.75
4.5°C	upper bound	3.92	2.59	2.98	4.43

TABLE 11.4: Global-mean Sea Level Changes (cm.) over 1990-2100 for Different Emissions Scenarios (IS92a, c, d and f), Different Values of the Climate Sensitivity (ΔT_{2x} = 1.5, 2.5 and 4.5°C) and using Both Lower and Upper Bound Estimates for the Projected CO_2 Concentration Changes. Radiative Forcing Changes Due to Gases other than CO_2 are Best Guess Values from Wigley and Raper (1992).

| ΔT_{2x} | CO_2 projection | IPCC Emissions Scenario | | | |
		IS92a	IS92c	IS92d	IS92e
1.5°C	lower bound	12	5	7	15
1.5°C	upper bound	16	8	10	19
2.5°C	lower bound	42	29	33	47
2.5°C	upper bound	50	37	40	55
4.5°C	lower bound	81	62	68	87
4.5°C	upper bound	93	74	79	99

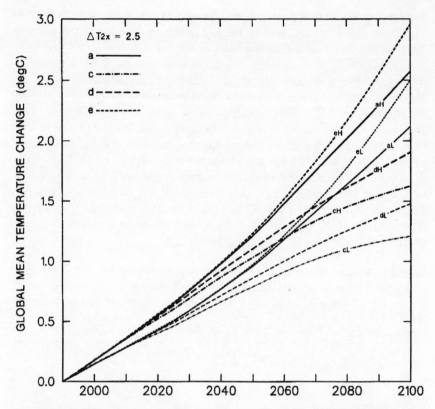

FIGURE 11.5: The effect of emissions scenario on carbon-cycle-related uncertainties in future global-mean temperature change. Upper, bolder curves use the upper-bound CO_2 projections (marked H, corresponding to "high" in Fig. 11.4), while lower, thinner curves use the lower-bound CO_2 projections (marked L, corresponding to "low" in Fig. 11.4). Up to around 2040, the effect of carbon cycle uncertainties outweighs the scenario influence (note the distinct separation of the two sets of curves). By 2100, scenario-to-scenario differences become far more important than carbon cycle uncertainty effects. (All temperature projections use a climate sensitivity of $\Delta T_{2x} = 2.5°C$).

Carbon cycle-related temperature uncertainties depend on the chosen value of the climate sensitivity, since, although the radiative forcing differential is fixed for any given scenario, the climate response to this differential increases as ΔT_{2x} increases. To illustrate this sensitivity to ΔT_{2x}, Fig. 11.6 shows the IS92a results for three values of ΔT_{2x} (1.5, 2.5 and 4.5°C) spanning the accepted range of uncertainty in this quantity. The corresponding global-mean temperature uncertainty ranges in the year 2100 are 0.31°C ($\Delta T_{2x} = 1.5°C$), 0.45°C ($\Delta T_{2x} = 2.5°C$) and 0.68°C ($\Delta T_{2x} = 4.5°C$). Similar uncertainty results are obtained for the other scenarios.

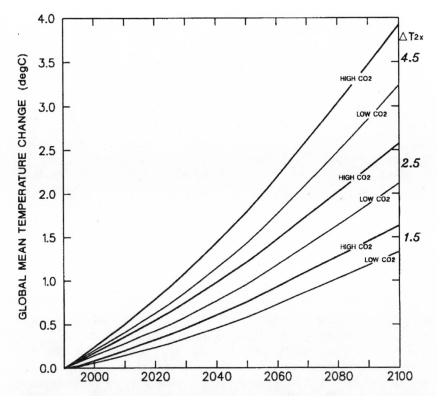

FIGURE. 11.6: The effect of climate sensitivity on carbon-cycle-related uncertainties in global-mean temperature for scenario IS92a. Upper, bolder curves use the upper-bound CO_2 projections ("high" in Fig. 11.4), while lower, thinner curves use the lower-bound CO_2 projections ("low" in Fig. 11.4). Climate sensitivities (ΔT_{2x}) are indicated on the right of the Figure. Climate sensitivity uncertainties are always more important than carbon cycle uncertainties.

Figures 11.5 and 11.6 allow one to compare the uncertainty arising from the carbon cycle model calculations (i.e., the difference between the high and low concentration estimates for each scenario) with the uncertainties arising from uncertainties in future emissions and ΔT_{2x}. The effect of emissions uncertainties is strongly time-dependent, being relatively small out to around 2040 and increasing rapidly after that date. Up to this time, carbon cycle uncertainties are more important than emissions uncertainties (Fig. 11.5). By 2100, however, emissions uncertainties are much more important. For $\Delta T_{2x} = 2.5°C$ and best estimates of the future concentration changes of all gases, emissions uncertainties give an uncertainty range of 1.3°C for global-mean temperature (the values for $\Delta T_{2x} = 1.5$ and 4.5°C are 0.9°C and 1.8°C, respectively).

Uncertainties in global-mean temperature arising from ΔT_{2x} uncertainties show a similar time dependence to those arising from carbon cycle uncertainties. ΔT_{2x} uncertainties are uniformly larger (see Fig. 11.6). By the year 2100, the uncertainty range varies from around 1.5°C for scenario c to around 2.4°C for scenario e, compared with uncertainty ranges of 0.3-0.7°C due to carbon cycle uncertainties (similar for all scenarios).

Thus, while it is clearly important to try to reduce carbon cycle modelling uncertainties, these uncertainties have noticeably smaller climate consequences than those arising from uncertainties in ΔT_{2x}. On the other hand, at least for the next 50 years, carbon-cycle-related climate change uncertainties are greater than those due to emissions uncertainties. They are also far larger than uncertainties arising from inadequacies in our understanding of the other gas cycles (see the accompanying chapter by Wigley).

Conclusions

Carbon dioxide is the most important contributor to both past and future anthropogenic climate change. Yet our ability to reliably predict future concentrations, or to fully explain past changes is still inadequate. These deficiencies are exemplified by the missing carbon problem: with best guess estimates of the rate of build up of atmospheric mass, the sizes of sources due to industrial activity and land-use change, and the strength of the sink of CO_2 into the ocean, there remains an imbalance – the so-called missing sink. There are a number of possible candidates for this missing sink, and a considerable incentive to explain it better. Its determination and quantification would lead to a considerable reduction in the uncertainties currently surrounding future concentration projections.

By using a carbon cycle model designed specifically to explore the concentration implications of uncertainties in the components of the carbon dioxide budget, I have quantified the ranges of likely concentration changes under the latest IPCC CO_2 emissions scenarios. Concentration levels for the different IPCC emissions scenarios in the year 2100 are:

IS92a; low	650ppmv; high	765ppmv;	difference	115ppmv
IS92c; low	436ppmv; high	505ppmv;	difference	69ppmv
IS92d; low	504ppmv; high	587ppmv;	difference	83ppmv
IS92e; low	875ppmv; high	1026ppmv;	difference	151ppmv

When converted to radiative forcing uncertainties, these concentration differences are equivalent to: IS92a, 1.02 W/m²; IS92c, 0.93 W/m²; IS92d, 0.96 W/m²; IS92e, 1.00 W/m².

The attendant uncertainties in global-mean temperature are large, although their precise values depend on the emissions scenario (slightly) and on the assumed value of the climate sensitivity, ΔT_{2x}. The 2100 temperature uncertainty values for scenario a, 0.31°C for $\Delta T_{2x} = 1.5$°C, 0.45°C for $\Delta T_{2x} = 2.5$°C and 0.68°C for $\Delta T_{2x} = 4.5$°C, are representative of all scenarios. These uncertainties are, however, much less than those due to uncertainties in ΔT_{2x}, viz. ranges of 2.2°C (IS92a), 1.5°C (IS92c), 1.7°C (IS92d) and 2.4°C (IS92e). If expressed in percentage terms, using (high-minus-low/2)/ (average of high and low), the uncertainty in global-mean temperature change over 1990-2100 due to carbon cycle uncertainties is ±8-16%, and that due to uncertainties in ΔT_{2x} is ±40-47%. Similar values apply to temperature changes over shorter periods.

The relative magnitudes of carbon-cycle- and emissions-related uncertainties in global-mean temperature vary according to time, largely because emissions uncertainties increase as one moves further into the future, while carbon cycle uncertainties (at least in percentage terms) remain relatively constant. Initially, carbon-cycle-related uncertainties are far more important. Beyond about 2040, however, emissions-related uncertainties become more important, increasingly so as time goes by. By the year 2100, uncertainties in global-mean temperature arising from CO_2 emissions uncertainties are about three times larger than those due to carbon cycle uncertainties, viz. ranges of 0.9°C ($\Delta T_{2x} = 1.5$°C), 1.3°C ($\Delta T_{2x} = 2.5$°C) and 1.8°C ($\Delta T_{2x} = 4.5$°C).

Is it possible to draw policy implications from, or assign research priorities on the basis of these results? Priorities should depend, not only on the absolute or relative uncertainties themselves, but also on their potential for reduction, the time required for such reduction and the cost of the research effort needed to achieve reduction.

Climate sensitivity leads to the biggest uncertainties, but the immediate potential for reduction of these uncertainties is, in my opinion, limited. The stated uncertainty range for ΔT_{2x} has been unchanged at 1.5-4.5°C for many years, in spite of a massive increase in the number of people engaged in climate research over the past 10-20 years and in spite of some major advances in climate modelling. To be fair, the subjectively-estimated confidence interval that this range represents (which is rarely stated in reviews of the subject) has probably narrowed. Furthermore, we now have a much better understanding of the climate system than we did 10 years ago – for example, we now know that the main factor leading to ΔT_{2x} uncertainties is cloud feedback (Cess et al., 1990). On the other hand, although there are research programmes underway that hope to reduce uncertainties in ΔT_{2x}, we do not know how rapidly this goal might be approached. It is not at all clear, for example, how a better understanding of cloud processes would significantly improve estimates of ΔT_{2x}, since the key problem of scaling up small-scale process results to the resolution scale of General Circulation Models, and larger, is not directly addressed by these studies. A cloud model that performs best at the small scale need not be the best at simulating large-scale changes, and validating the various small-to-large-scale parameterizations in the context of climate *change* is thwarted at the outset by the lack of suitable validation data. Empirical analyses of past temperature changes allow ΔT_{2x} to be estimated, and the range of likely values will decrease as time goes by and more data become available – but this reduction in uncertainty will be slow (Wigley and Raper, 1991).

Emissions scenario uncertainties arise mainly from uncertainties in: future population growth; changes in energy use per capita; and changes in emissions per unit of energy production. Improvements in energy-economics modelling will lead to some reduction in future emissions uncertainties. However, the problems of population projection and of any form of economic projection in the developing world are enormous and enormously complex. Furthermore, emissions uncertainties are relatively unimportant out to the middle of the next century.

Of all three sources of uncertainty, those related to carbon cycle modelling can probably be reduced most for the least effort and smallest cost; not least because the research effort in this area is still limited to a small community. The key issues here are to improve the ocean modelling component and to better identify the components of the missing sink.

Economic analyses (e.g., Manne and Richels, 1991; Peck and Teisberg, 1992) show that reducing uncertainty has a high economic value – research directed to this end appears to be far more cost effective than, for example, attempts to reduce greenhouse gas emissions beyond the ill-defined "no regrets" stage. Of course, if a large research effort results in little reduction in uncertainty, then research would *not* be cost effective. A full economic assessment of the value of research requires a probabilistic assessment of the degree of reduction of uncertainty and the costs of achieving such reductions. To my knowledge, no one has ever attempted to do this. Certainly, little or no work has been carried out to assess the relative cost effectiveness of different research activities in the greenhouse context. To do this convincingly requires more than just an assessment of the probability for success. My own subjective view is that carbon cycle research is one of the best avenues (in the sense that there is a high probability that uncertainties can be reduced in a relatively short time), but there is considerable scope for putting this opinion on a firmer basis.

References

Briffa, K. R. 1992. "Increasing productivity of 'natural growth' conifers in Europe over the last century," in, *Tree Rings and Environment, Proceedings of the International Dendrochronological Symposium*, Ystad, Sweden (T.S. Bartholin, B.E. Berglund, D. Eckstein and F.H. Schweingruber, Eds.). Lundqua Report 34, Lund University, pp. 64-71.

Cess, R. D. et al. 1990. "Interpretation of cloud-climate feedback produced by 14 atmospheric general circulation models." *Science* 245, 513-516.

Enting, I. G. and Mansbridge, J. V. 1991. "Latitudinal distribution of sources and sinks of CO_2: results of an inversion study." *Tellus* 43B, 156-170.

Folland, C. K., Karl, T. R. and Vinnikov, K. Ya. 1990. "Observed climate variations and change," in, *Climate Change: The IPCC Scientific Assessment* (J. T. Houghton, G.J. Jenkins and J.J. Ephraums, Eds.), Cambridge University Press, pp. 195-238.

Folland, C. K., Karl, T.R ., Nicholls, N., Nyenzi, B.S., Parker, D. E. and Vinnikov, K.Ya. 1992. "Observed climate variability and change," in, *Climate Change 1992: The Supplementary Report to the IPCC Scientific Assessment* (J.T. Houghton, B.A. Callander and S. K. Varney, Eds.). Cambridge University Press, pp. 135-170.

Harvey, L.D.D. 1989b. "Effect of model structure on the response of terrestrial biosphere models to CO_2 and temperature increases." *Global Biogeochemical Cycles* 3, 137-153.

Harvey, L. D. D. 1989a. "Managing atmospheric CO_2." *Climatic Change* 15, 343-381.

Houghton, J.T., Jenkins, G.J. and Ephraums, J.J. (Eds.), 1990: *Climate Change: The IPCC Scientific Assessment*. Cambridge University Press.

Houghton, R. A. 1991. "The role of forests in affecting the greenhouse gas composition of the atmosphere," in, *Global Climate Change and Life on Earth* (R.L. Wyman, Ed.). New York: Chapman and Hall, pp. 43-55.

Hoyt, D. V., Kyle, H. L., Hickey, J. R. and Maschhoff, R.H. 1992. The Nimbus 7 solar total irradiance: A new algorithm for its derivation. *Journal of Geophysical Research* 97, 51-63.

Innes, J. L. 1991. "High-altitude and high-latitude tree growth in relation to past, present and future global climate change." *The Holocene* 1, 168-173.

Kauppi, P. E., Mielikinen, K. and Kuusela, K. 1992. "Biomass and carbon budget of European forests, 1971-1990." *Science* 256, 70-74.

Keeling, C. D. 1991. "CO_2 emissions – historical record, global,"in, *Trends 91: A Compendium of Data on Global Change* (T.A. Boden, R.J. Sepanski and F.W. Stoss, Eds.), ORNL/CDIAC-46. Carbon Dioxide Information Analysis Center, Oak Ridge, TN, pp. 382-385.

Kelly, P. M. and Wigley, T. M. L. 1992. "Solar cycle length, greenhouse forcing and global climate." *Nature* 360, 328-330.

Kiehl, J. T. and Briegleb, B. P. 1993. "The relative role of sulfate aerosols and greenhouse gases in climate forcing." *Science* 260, 311–314.

Leggett, J., Pepper, W. J. and Swart, R. J. 1992. "Emissions scenarios for IPCC: An update," in, *Climate Change 1992: The Supplementary Report to the IPCC Scientific Assessment* (J.T. Houghton, B.A. Callander and S.K. Varney, Eds.). Cambridge University Press, pp. 69-96.

Manne, A. S. and Richels, R .G. 1991. "Buying greenhouse insurance." *Energy Policy* 19, 543-552.

Marland, G. and Boden, T. A. 1991. "CO₂ emissions-modern record, global," in, *Trends 91: A Compendium of Data on Global Change* (T.A. Boden, R.J. Sepanski and F.W. Stoss, Eds.), ORNL/CDIAC-46. Carbon Dioxide Information Analysis Center, Oak Ridge, TN, pp. 386-389.

Peck, S. C. and Teisberg, T. J. 1992. "Global warming uncertainties and the value of information: An analysis using CETA." (Draft manuscript, EPRI Environmental Sciences Dept.)

Ramaswamy, V., Schwarzkopf, M. D. and Shine, K. P. 1992. "Radiative forcing of climate from halocarbon-induced global stratospheric ozone loss." *Nature* 355, 810-812.

Sarmiento, J. L. and Sundquist, E. T. 1992. "Revised budget for the oceanic uptake of anthropogenic carbon dioxide." *Nature* 356, 589-593.

Shine, K. P., Derwent, R. G., Wuebbles, D. J. and Morcrette, J. J. 1990. "Radiative forcing of climate," in, *Climate Change: The IPCC Scientific Assessment* (J. T. Houghton, G. J. Jenkins and J. J. Ephraums, Eds.). Cambridge: Cambridge University Press, pp. 41-68.

Tans, P. P., Fung, I. Y. and Takahashi, T. 1990. "Observational constraints on the global atmospheric CO₂ budget." *Science* 247, 1431-1438.

Wang, W.-C., Dudek, M. P., Liang, X.-Z. and Kiehl, J. T. 1991. "Inadequacy of effective CO₂ as a proxy in simulating the greenhouse effect of other radiatively active gases." *Nature* 350, 573-577.

Watson, R. T., Rodhe, H., Oeschger, H. and Siegenthaler, U. 1990. "Greenhouse gases and aerosols," in, *Climate Change: The IPCC Scientific Assessment* (J.T. Houghton, G.J. Jenkins and J.J. Ephraums, Eds.). Cambridge: Cambridge University Press, pp. 1-40.

Wigley, T. M. L. 1991. "A simple inverse carbon cycle model." *Global Biogeochemical Cycles* 5, 373-382.

Wigley, T. M. L. 1993. "Balancing the carbon budget: implications for projections of future carbon dioxide concentration changes." *Tellus*.

Wigley, T. M. L. and Barnett, T. P. 1990. "Detection of the greenhouse effect in the observations," in, *Climate Change: The IPCC Scientific Assessment* (J.T. Houghton, G.J. Jenkins and J.J. Ephraums, Eds.). Cambridge: Cambridge University Press, pp. 239-256.

Wigley, T. M. L., Briffa, K. R. and Jones, P. D. 1984. "Predicting plant productivity and water resources." *Nature* 312, 102-103.

Wigley, T. M. L. and Kelly, P. M. 1990. "Holocene climatic change, 14C wiggles and variations in solar irradiance." *Philosophical Transactions of the Royal Society* A330, 547-560.

Wigley, T. M. L. and Raper, S.C.B. 1991. "Detection of the enhanced greenhouse effect on climate," in, *Climate Change: Science, Impacts and Policy* (J. Jäger and H. L. Ferguson, Eds.). Cambridge University Press, pp. 231-242.

Wigley, T. M. L. and Raper, S. C. B. 1992. "Implications for climate and sea level of revised IPCC emissions scenarios." *Nature* 357, 293-300.

Zöttl, H.W., Hüttl, R.F., Fink, S., Tomlinson, C.H. and Wisniewski, J., 1989: Nutritional disturbances and histological changes in declining forests. *Water, Air and Soil Pollution* 48, 87-109.

Support from the U.S. Dept. of Energy (Grant No. DE-FG02-86ER60397) is gratefully acknowledged. This work was carried out while the author was at the Climatic Research Unit, University of East Anglia, Norwich, U.K.

12

The Contribution From Emissions of Different Gases to the Enhanced Greenhouse Effect

T. M. L. Wigley

Abstract

This chapter quantifies the relative importance of different anthropogenic gases in modifying past and future climate, together with the associated uncertainties. Past radiative forcing changes or future changes under the 1992 IPCC emissions scenarios are used as a basis for comparison. For the past, CO_2 is by far the most important contributor to forcing changes, followed by SO_2-derived sulphate aerosols and CH_4. The largest uncertainties are those associated with aerosol forcing. For the future, the largest forcing contributions come from CO_2, followed by aerosols, CH_4 and then either N_2O or the halocarbons (depending on scenario). Future forcing uncertainties arising from the following factors are quantified: emissions uncertainties; possible changes in lifetimes (for CH_4 and hydrogenated halocarbons); stratospheric ozone feedback (for halocarbons); and global-warming-related emissions increases (for CH_4). The most important absolute uncertainties are, in order of magnitude, those associated with CO_2 changes, aerosols and methane. Research priorities should be determined by the absolute magnitudes of uncertainty (which requires the use of a common yardstick such as radiative forcing) and by the potential for reductions in uncertainty. On this basis, carbon cycle and aerosol research (including the issue of determining the spatial patterns of aerosol forcing) are judged of highest priority, followed by the atmospheric chemistry of methane and climate-related methane emissions changes. Other important issues, but of lower priority, are the current N_2O budget, possible climate-related changes in N_2O emissions, stratospheric water vapour changes arising from methane emissions and the indirect radiative forcing from halocarbons arising through stratospheric ozone depletion.

Introduction

The main purpose of this chapter is to compare the different contributions, both past and future, that humankind has made to perturbing the atmosphere's radiative balance. We have, and will continue, to perturb both the balance of outgoing long wave radiation and the balance of incoming short wave radiation. In the former category, human activities since pre-industrial times have caused a substantial enhancement of the greenhouse effect, a process involving the absorption of outgoing long-wave radiation which leads to a warming of the lower atmosphere. In the latter category, because the atmosphere's short-wave radiative balance is affected by the presence of small particles (aerosols) produced by the oxidation of sulphur compounds, anthropogenic emissions of sulphur dioxide (SO_2) have also caused a perturbation of the overall balance.

The greenhouse gases considered in this paper are, in order of importance: carbon dioxide (CO_2), methane (CH_4), nitrous oxide (N_2O) and the halocarbons. Tropospheric ozone (O_3) is also a greenhouse gas, but it is not considered here because its contribution to the greenhouse effect is yet to be reliably quantified (Shine et al., 1990; Isaksen et al., 1992a).

The simplest way to quantify and compare these various factors is to use their individual global-mean net radiative perturbations at the top of the troposphere, ΔQ in W/m^2: the larger the value of ΔQ, the larger the potential impact on climate. Using global-mean ΔQ is a considerable simplification. It hides important spatial forcing details, which vary markedly from gas to gas. For SO_2 and the halocarbons, the pattern of forcing varies greatly both seasonally and geographically (for SO_2, see Charlson et al., 1991; for halocarbons, see Ramaswamy et al., 1992, and Isaksen et al., 1992a), and even the pattern of CO_2 forcing is far from uniform. The use of a single, global-mean indicator, while a somewhat deceptive over-simplification is, nevertheless, of considerable value because the pattern of climate change tends to be relatively insensitive to the pattern of forcing (Manabe and Wetherald, 1980; Hansen et al., 1984; Wang et al., 1991).

The value of ΔQ for a particular gas is determined by changes in its prevailing global-mean concentration, C. These, in turn, are determined by its history of emissions changes. For the past, we have reasonably accurate information regarding past concentration changes (Watson et al., 1990, 1992). These are readily converted to histories of ΔQ for each gas, albeit with some additional uncertainty. For the future, concentration changes, $\Delta C(t)$, can be estimated from scenarios of possible future emissions, $\Delta E(t)$. For any given emissions scenario, this introduces further uncertainties because the gas cycle models used to relate ΔE and ΔC are still subject to considerable uncertainty.

In the following, observed and model-based concentration data are used, together with the most recent information relating concentrations to radiative forcing (Shine et al., 1990), to estimate the individual contributions of the different gases to the changing radiative balance of the atmosphere. Ranges of uncertainty in each of these estimates are also estimated. For the future, all results are based on the 1992 IPCC emissions scenarios (Leggett et al., 1992), IS92a–f. I begin with a summary of 1990 conditions, then consider each gas separately (but lumping the halocarbons into a single group), compare their relative importance, and finally discuss the uncertainties with a view to assessing priorities for research.

The 1990 Situation

The standard starting point for calculations is the nominal pre-industrial date of 1765. At that time, concentrations of CO_2, CH_4 and N_2O were around 279ppmv, 790ppbv and 285ppbv respectively (Watson et al., 1990; Shine et al., 1990); but note that more recent data suggests a lower pre-industrial level of 260–270ppbv for N_2O (Leuenberger and Siegenthaler, 1992). Halocarbon concentrations were effectively zero until around 1950. 1990 (mid-year) concentration levels are listed in Table 12.1. Table 12.1 also shows 1990 emissions levels, the lifetimes of the gases, and their radiative forcing sensitivities (i.e., ΔQ per unit concentration change). Concentrations, lifetimes and radiative forcing sensitivities differ by many orders of magnitude. The halocarbons, by far the least abundant gases, have the highest sensitivities, bringing their total radiative effects to date relatively close to those for the other, more abundant gases (see Fig. 12.1).

The list of halocarbons shown in Table 12.1 is not exhaustive, but it probably accounts for the major part (>95%) of the halocarbon forcing to 1990. It includes all the gases given in the IPCC92 emissions scenarios bar HCFC225 (which contributes zero to date, and negligibly in the future). Some of the halocarbons in Table 12.1 were not covered by IPCC92 (viz. CFC13, CFC14, CFC116, chloroform and methylene chloride) – they are included here because of their current non-negligible abundance and/or potential future importance.

For the halocarbons, radiative sensitivities are constants (with values probably uncertain by ±10–20%). For CO_2, CH_4 and N_2O, ΔQ depends nonlinearly on ΔC, so $\partial Q / \partial C$ must depend on the concentration level,

TABLE 12.1: 1990 Values For Greenhouse Gas Emissions, Concentrations, Lifetimes and Radiative Forcing Sensitivities ($\partial Q/\partial C$). Concentrations are Global-, Annual-Means in ppbv, Lifetimes are in Years and $\partial Q/\partial C$ Values are in W/m^2 per ppbv. Concentration and Lifetime Data are From IPCC Unless Otherwise Stated. Gases Marked * Were Not Included in the IPCC92 Emissions Scenarios, but are Included in the Calculations Presented here.

Gas Name	Formula	Emissions[1]	Conc.	Lifetime	$\partial Q/\partial C$[2]
Carbon Dioxide	(CO$_2$)	7.4	354000.	(100)	.000018
Methane[3]	(CH$_4$)	506.	1717.	10.5	.000376
Nitrous Oxide	(N$_2$O)	12.9	310.	132.	.003770
CFC11	(CCl$_3$F)	298.	.280	55.	.220
CFC12	(CCl$_2$F$_2$)	363.	.484	116.	.280
*CFC13[4,5]	(C$_2$Cl$_3$F$_3$)	2.	.003	400.	.345
*CFC14[6]	(CF$_4$)	20.	.084	10000.	.100
CFC113	(C$_2$Cl$_3$F$_3$)	147.	.070	110.	.279
CFC114	(C$_2$Cl$_2$F$_4$)	13.	.015	220.	.323
CFC115	(C$_2$ClF$_5$)	7.	.005	550.	.257
*CFC116[6]	(C$_2$F$_6$)	1.3	.004	10000.	.300
Carbontetrachloride	(CCl$_4$)	119.	.108	47.	.101
*Chloroform[4,5]	(CHCl$_3$)	300.	.010	.7	.094
*Methylene Chloride[4,5]	(CH$_2$Cl$_2$)	600.	.026	.6	.046
Methyl Chloroform	(CH$_3$CCl$_3$)	738.	.158	6.1	.048
Ha1211[7]	(CF$_2$ClBr)	3.	.002	11.	.230
Ha1301	(CF$_3$Br)	4.	.003	77.	.284
HCFC22	(CHClF$_2$)	138.	.115	15.8	.189
HCFC123	(CHCl$_2$CF$_3$)	.0	.0	1.7	.176
HCFC124	(CHClFCF$_3$)	.0	.0	6.9	.191
HCFC141b	(CH$_3$CCl$_2$F)	.0	.0	10.8	.136
HCFC142b	(CH$_3$CClF$_2$)	.0	.0	22.4	.180
HFC125	(CHF$_2$CF$_3$)	.0	.0	40.5	.238
HFC134a	(CH$_2$FCF$_3$)	.0	.0	15.6	.169
HFC152a	(CH$_3$CHF$_2$)	.0	.0	1.8	.117

[1] Values are as given in the IPCC92 emissions scenarios, unless stated otherwise. Units are GtC/yr for CO$_2$, TgCH$_4$/yr for CH$_4$, TgN/yr for N$_2$O and Ktonne/yr for all halocarbons.

[2] Note that $\partial Q/\partial C$ depends on C for CO$_2$, CH$_4$ and N$_2$O.

[3] Lifetime includes effect of soil sink. $\partial Q/\partial C$ includes stratospheric water vapour effect as given by Shine et al. (1990).

[4] Emissions estimate based on concentration data.

[5] 1990 concentrations from S.E. Penkett (personal communication): CFC13, 3.5pptv; chloroform, 10pptv; methylene chloride, 26pptv.

[6] Data from or based on Isaksen et al. (1992b).

[7] $\partial Q/\partial C$ estimated from chemical structure.

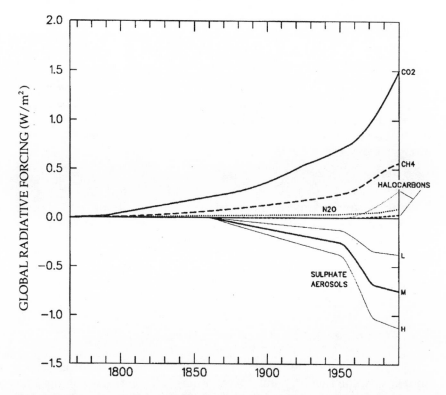

FIGURE 12.1: (From Wigley, 1993a.) Global-mean radiative forcing since pre-industrial times (1765) arising from man-made emissions of CO_2 (upper full line), CH_4 (long dashes), N_2O (dots), halocarbons (thin dots or short dashes) and sulphate aerosols (full lines below zero). N_2O values assume an initial concentration of 285ppmv (following IPCC). For an initial value of 265ppmv (cf. Leuenberger and Siegenthaler, 1992) the ΔQ values for N_2O are about double those shown here. For halocarbons, the upper curve corresponds to the IPCC estimate given by Shine et al. (1990), while the lower estimate accounts for the negative forcing effect of stratospheric ozone depletion following Ramaswamy et al. (1992). The aerosol forcing values shown are low, mid and high estimates from Wigley and Raper (1992).

decreasing as C increases. Lifetimes of the non-hydrogenated halo-carbons are effectively constant. For the HCFCs and HFCs, since they are chemically removed by oxidation in the troposphere, their lifetimes will vary as atmospheric composition changes. For some future scenarios, these lifetimes may change noticeably, but the overall importance of these changes is very small. The changes should parallel those of CH_4, since the lifetimes of both hydrogenated halocarbons and CH_4 depend mainly on the same factor, viz. the prevailing concentration of the OH radical. Lifetimes of these gases may therefore increase by up to 20% by the year 2100 (see below).

Carbon Dioxide

Carbon dioxide is discussed in detail in the accompanying chapter by Wigley (1993a). Here, I simply summarize the main results of that chapter, see Fig. 12.2. This figure shows the IPCC92 emissions scenarios, implied (best-guess) concentration projections, and consequent radiative forcing changes from 1990.

Methane

Future methane concentration changes will be determined by three factors: direct emissions changes due to human activities; changes in atmospheric chemistry (which will affect the lifetime of methane); and feedback effects, which may alter emissions from the biosphere.

Present knowledge of methane emissions is uncertain, but different approaches to estimating the current net emissions level give similar results. From mass balance considerations, using

$$dC/dt = E/2.75 - C/\tau - C/\tau_s$$

(where 2.75 converts atmospheric mass in $TgCH_4/yr$ to lower tropospheric concentration in ppbv, τ is the atmospheric chemical lifetime and τ_s is the effective lifetime of the soil sink), the 1990 emissions level is 450(480)520$TgCH_4/yr$ for assumed τ of 12.3(11.3)10.3yr (using C = 1717ppbv, dC/dt = 12ppbv/yr and τ_s = 150yr). From Watson et al. (1992), the best-guess value of E is 515$TgCH_4/yr$, with a range of 331–850$TgCH_4/yr$. The value assumed by Leggett et al. (1992) as a 1990 starting point for the IPCC92 emissions scenarios is 506$TgCH_4/yr$. The results given below use a 1990 value of 480$TgCH_4/yr$, consistent with the best guess lifetime of 11.3yr.

Figure 12.3 shows the IPCC92 emissions scenarios, together with best-guess concentration projections from Osborn and Wigley (1993) and corresponding radiative forcing changes. The concentration projections account for lifetime changes (see Table 12.2), which in turn are determined by the emissions of the gases that are the primary determinants of OH concentration, viz. CH_4, carbon monoxide (CO), nitrogen oxides (NO_x) and non-methane hydrocarbons (NMHCs). Emissions used for CO, NO_x and NMHCs are those given in the IPCC92 scenarios. Radiative forcing changes for methane are calculated using IPCC92 methods (Shine et al., 1990) – as is the case for all greenhouse gases in this review. Stratospheric water vapour effects are included as in Shine et al.

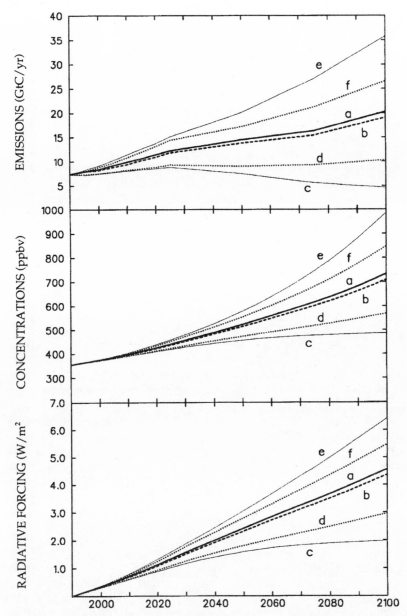

FIGURE 12.2: CO_2 emissions for the IPCC92 scenarios (IS92a–f) together with implied concentration and radiative forcing changes. Concentration changes are best guess values calculated using the carbon cycle model of Wigley (1993b). Radiative forcing values use these concentrations and the relationship $\Delta Q = 6.3\ln(C/C_O)$ recommended by Shine et al. (1990).

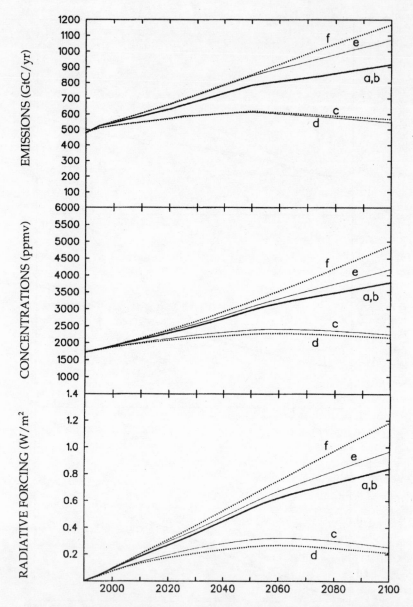

FIGURE 12.3: CH$_4$ emissions for the IPCC92 scenarios (IS92a–f) together with implied concentration and radiative forcing changes. Concentration changes are best guess values calculated using the model of Osborn and Wigley (1993). Radiative forcing values, which include the effects of stratospheric water vapour changes arising from methane oxidation, are based on expressions given in Shine et al. (1990).

TABLE 12.2: Methane Lifetime Changes (2100 Atmospheric Chemical Lifetime Divided By 1990 Lifetime) For the IPCC92 Emissions Scenarios, From Osborn and Wigley (1993).

Scenario	Low	Mid	High
IS92a,b	1.06	1.15	1.25
IS92c	1.02	1.05	1.11
IS92d	0.94	0.97	0.99
IS92e	1.03	1.09	1.16
IS92f	1.09	1.20	1.34

(1990), although it should be noted that there is considerable uncertainty surrounding this term (Isaksen et al., 1992a).

To give some idea of the importance of lifetime uncertainties and changes, Fig. 12.4 shows low, best-guess and high concentration projections for each scenario together with results assuming a constant lifetime of 11.3yr (from Osborn and Wigley, 1993). In terms of concentration change, the maximum uncertainty arising through lifetime uncertainties is roughly equal to the separation between the best-guess results for scenarios IS92e and IS92f. In terms of radiative forcing, this amounts to about $0.2W/m^2$ ($\pm 0.1W/m^2$) by 2100 (see Fig. 12.3), a relatively small effect.

The other main source of uncertainty for methane arises through possible feedbacks on biospheric emissions, where global-scale warming and associated climate changes lead to changes in these emissions. A rough estimate of the potential magnitude of this effect can be obtained from paleo-data. From ice-core concentration data, we know that methane levels were about 340ppbv at the peak of the last glacial period (around 20,000 years ago) and 790ppbv around the late 18th century. Lifetime estimates for these two periods have been made by numerous authors; Osborn and Wigley (1993) give values of 8.0yr and 8.3yr respectively. For a soil sink lifetime of 150 years, the mass balance equation implies natural emissions levels of $125TgCH_4yr$ at glacial maximum and $275TgCH_4/yr$ in the late 18th century. The corresponding global-mean temperature change was 4–5°C, somewhat more than the best-guess warming over 1990–2100 given by Wigley and Raper (1992), viz. 2.5°C.

For a 4–5°C warming, therefore, the warming-related feedback effect could lead to additional emissions of up to $150TgCH_4/yr$ by 2100, similar to the difference between scenarios IS92a and IS92e. This is likely to be an upper bound, because the temperature, land character and precipitation changes over the past 20,000 years substantially exceed those expected over the next century. Indeed, methane feedback effects are such that negative and positive feedbacks could largely cancel each other leading to a much smaller net effect (see, e.g., many of the papers

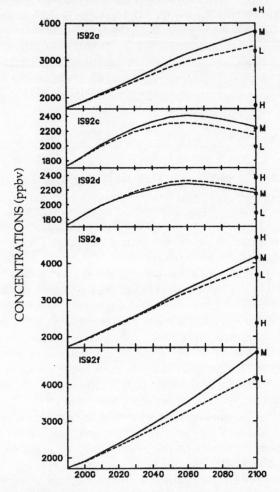

FIGURE 12.4: Methane concentration uncertainties and the effect of lifetime variations on concentration projections for the IPCC92 emissions scenarios (IS92a–f). Full lines show results for variable lifetime, while dashed lines show results if the atmospheric chemical lifetime is assumed to be constant at its 1990 level of 11.3yr. Lifetime changes, which depend on the emissions of CO, NO_x and NMHCs (as given in the IPCC92 scenarios) and on the prevailing concentrations of CH_4, were calculated using the model of Osborn and Wigley (1993). In all except IS92d, lifetime is projected to increase, leading to concentration levels above the constant-lifetime projections. Dots along the right-hand side (year 2100) show low and high projections allowing for uncertainties in the current lifetime and future lifetime changes, based on the Osborn/Wigley model. Results for IS92b are not shown because they are virtually identical to those for IS92a.

in Rogers and Whitman, 1991). The radiative forcing uncertainty by 2100 due to feedback, therefore, ranges between around zero and the difference between the IS92a and IS92e ΔQ values, i.e. 0–0.2W/m². As for the lifetime variation influence, this is also a relatively small effect.

The longer time scale consequences of feedbacks, may, however, be more substantial because methane reservoirs not noticeably affected by glacial/interglacial climate changes could be influenced substantially when future warming surpasses anything experienced in the "recent" (past 10^6 yr) geological past.

Nitrous Oxide

Future nitrous oxide emissions under the IPCC92 scenarios, together with concentration changes and equivalent radiative forcing changes are shown in Fig. 12.5. Concentration changes shown here are based on the mass balance equation

$$dC/dt = E/4.81 - C/\tau$$

where the factor 4.81 converts atmospheric mass in TgN to lower troposphere concentration in ppbv, and where 132yr has been used as the lifetime (following IPCC92; down from the 150yr value used by IPCC in 1990).

Considerable uncertainties attend the projection of N_2O concentration changes. While scientifically important, these uncertainties are of little immediate importance to the issue of future climatic change since, to 2100, the maximum contribution of N_2O to radiative forcing changes is small (around 0.3W/m²). On very long time scales, however, N_2O becomes more important because of its long lifetime and uncertainties regarding the budget of the gas (the balance between sources, sinks and atmospheric buildup).

Budget uncertainties are epitomized by the very different results obtained for net 1990 emissions using different methods. From the mass balance equation and current concentration data, the 1990 net emissions level is 13.2(14.7)16.7TgN/yr for t = 152(132)112yr. Based on estimates of individual sources, Watson et al. (1992) give lower and upper bounds of 5.2 and 16.1TgN/yr (mean, 10.6TgN/yr). The scenarios developed by Leggett et al. (1992) use 12.9TgN/yr as the 1990 level. While these ranges overlap, they show large differences in their central values: the difference between 10.6 and 14.7TgN/yr is comparable to the total change projected between 1990 and 2100 under scenarios IS92e and f (viz. about 6TgN/yr). The chosen level of emissions in 1990 does noticeably affect the concentration projections for N_2O. For example, using the value 14.7TgN/yr that is consistent with the mass balance equation (rather

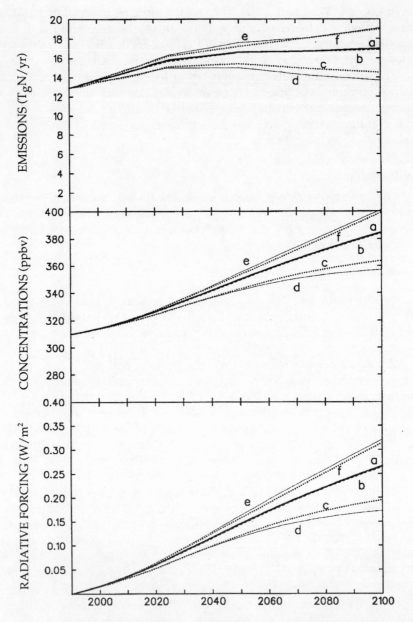

FIGURE 12.5: N$_2$O emissions for the IPCC92 scenarios (IS92a–f) together with implied concentration and radiative forcing changes. Concentration changes are based on a simple mass balance model with constant lifetime of 132yr (the value recommended by Isaksen et al., 1992a). Radiative forcing values were calculated using the expression given by Shine et al. (1990).

than 12.9TgN/yr) and assuming that the *changes* from 1990 are as given in IS92a, leads to a 2100 concentration of 413ppbv (rather than 385ppbv) and a radiative forcing change over 1990-2100 of 0.36W/m² (rather than 0.27W/m²).

Apart from this initial value uncertainty, there are large uncertainties in the projected changes in N_2O emissions. For example, possible natural emissions changes and feedbacks on these changes due to global warming were not considered in the IPCC scenarios – largely because these effects are not quantifiable at present. We know, however, that natural N_2O emissions have varied markedly in the past (at least on long time scales). The concentration level at the last glacial maximum was only 180-200ppbv (Leuenberger and Siegenthaler, 1992), compared with 260-270ppbv (op. cit.) or 285ppbv (IPCC) in the late eighteenth century, implying either a similar proportional change in natural emissions (or, far less likely, a large change in lifetime).

Uncertainties in future emissions also imply proportionally similar uncertainties in future concentration levels; although, because of the long lifetime of N_2O, these changes will only be fully manifest after times of order centuries. The eventual magnitude of N_2O forcing could, therefore, be much larger than any change that might occur to 2100. For instance, if N_2O emissions were to double, the concentration level would eventually double (compared to the maximum change of only +30% by 2100 under the IPCC92 scenarios). This would give a radiative forcing change of around 1W/m², making N_2O potentially more important than CH_4 on time scales greater than a century.

Halocarbons

Because of the controls on halocarbon production internationally legislated under the Montreal Protocol, the future enhanced greenhouse effect of halocarbons is relatively small – in spite of the fact that some halocarbons are more than 10,000 times stronger greenhouse gases than CO_2, mass for mass (see Table 12.1). The strengthening of the Protocol between 1990 and 1992 means that the likely total halocarbon ΔQ based on the IPCC92 scenarios is much less than that based on the IPCC90 scenarios. Compared with ΔQ estimates given by IPCC in 1990, the latest estimates (Wigley and Raper, 1992) are even smaller because of the incorporation of the radiative effects of stratospheric ozone changes in the most recent calculations.

The six IPCC92 scenarios, IS92a–f, contain only three distinct halocarbon emissions scenarios: IS92a, c and f are the same, as are IS92d and e. Scenarios d and e correspond quite closely to the most recently proposed (Copenhagen) amendments to the Montreal Protocol (which were

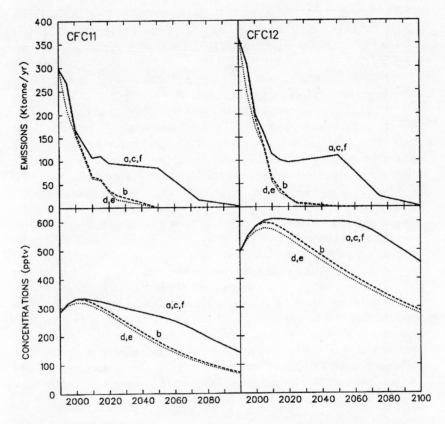

FIGURE 12.6: IPCC92 emissions scenarios and implied concentration changes for CFC11 (left panels) and CFC12 (right panels). Concentration changes were calculated using simple mass balance models with lifetimes of 55yr (CFC11) and 116yr (CFC12), the values recommended by Isaksen et al. (1992a).

made after the IPCC92 report was published). Scenario b represents a mix of IS92a,c,f details with those of IS92d,e (see Figs. 12.6 and 12.7).

In the IPCC92 scenarios developed by Leggett et al. (1992), 18 individual halocarbons were considered. In the ΔQ assessment of Wigley and Raper (1992), one of these was omitted (HCFC225). This was mainly due to lack of suitable lifetime and radiative forcing information; but with reasonable estimates for these quantities the contribution of this gas to the overall forcing is negligible. Wigley and Raper also included CFC13, CFC14, CFC116, chloroform and methylene chloride in their assessments. They made certain assumptions for the future emissions of these gases (the assumptions for chloroform and methylene chloride were unfortunately given in reverse in their manuscript). Here we modify these

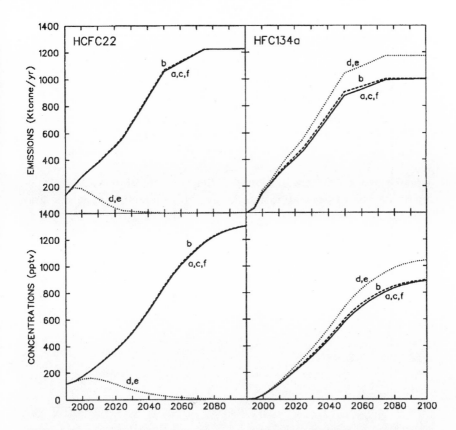

FIGURE 12.7: IPCC92 emissions scenarios and implied concentration changes for HCFC22 (left panels) and HFC134a (right panels). Concentration changes were calculated using simple mass balance models with lifetimes of 15.8yr (HCFC22) and 15.6yr (HFC134a), the values recommended by Isaksen et al. (1992).

assumptions for these gases slightly. For CFC13, we follow Wigley and Raper by assuming the ratio of CFC13 to CFC12 emissions remains constant. For CFC14 and CFC116 we assume constant emissions at 1990 levels estimated from data given by Isaksen et al. (1992b), viz. 20ktonne/ yr and 1.3ktonne/yr respectively (Wigley and Raper used 66ktonne/yr and 6ktonne/yr, values that now appear to be far too high). For chloroform, we also assume constant emissions at a newly estimated 1990 level of 300ktonne/yr (based on concentration data). For methylene chloride we assume constant emissions at 1100ktonne/yr (also based on concentration data).

The dominant halocarbons in terms of current concentrations are CFC11 and CFC12 (see Table 12.1). For future concentrations under the

Montreal Protocol, the dominant gases are likely to be HCFC22 and HFC134a, the primary substitute chemicals. In Figs. 12.6 and 12.7, we show IPCC92 emissions scenarios for these four gases, together with concentration projections based on mass balance equations with constant lifetimes. Although the lifetimes of HCFC22 and HFC134a are likely to increase in the future, by up to +20% using CH_4 as an analogue species (see Table 12.2), this is a relatively small and secondary factor, with a negligible effect in terms of radiative forcing changes.

Radiative forcing projections for the sum of all halocarbons are shown in Fig. 12.8. This Figure shows results with and without the effect of stratospheric ozone changes. As noted in IPCC92, chlorine- and bromine-containing halocarbons almost certainly cause depletion of lower stratospheric ozone, and this affects the radiative balance of the troposphere in a way that offsets the direct radiative effects of these chemicals.

Wigley and Raper (1992) introduced a simple method to account for this offsetting effect, similar to the concept of chlorine-loading potential that has been used to assess the ozone depletion effect of halocarbons. Based on the results of Ramaswamy et al. (1992), they assume that the total (global-mean) radiative forcing for a particular halocarbon (ΔQ_{TOT} in W/m²) due to a concentration change (ΔC in ppbv) is

$$\Delta Q_{TOT} = \Delta Q_{DIR} - \Delta Q_{OZ}$$

where ΔQ_{DIR} is the direct effect,

$$\Delta Q_{DIR} = a\Delta C = (\partial Q/\partial C)\Delta C$$

(with a $= \partial Q/\partial C$ values given in Table 1) and ΔQ_{OZ} is the ozone-change contribution

$$\Delta Q_{OZ} = 0.0762\, N_{Cl}\, \Delta C$$

Here, N_{Cl} is the number of chlorine atoms in the molecule. The factor 0.0762 was obtained by calibration against the results of Ramaswamy et al. (1992). For bromine-containing compounds, Wigley and Raper assumed one Br atom equivalent to five Cl atoms. Here we use a more realistic multiplying factor of 40 (although this change has only a very small effect on the final results).

Figure 12.8 illustrates a number of important points. First, it shows the substantial reduction in overall halocarbon ΔQ due to the strengthening of the Montreal Protocol between 1990 and 1992. This differential is considerably less, however, if stratospheric ozone effects are accounted for. Second, it shows that the effect of including ozone in the calculations dif-

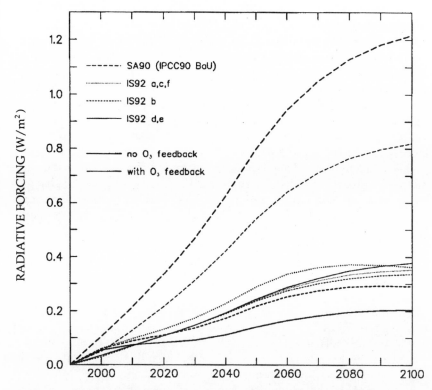

FIGURE 12.8: Total halocarbon forcing changes from 1990 for IPCC emissions scenarios SA90 (the 1990 IPCC "Business as Usual" scenario), IS92a,c,f, IS92b and IS92d,e. Bold lines or large dots show results with stratospheric ozone effects included, while thin lines or small dots show results with ozone effects neglected. Key: SA90, long dashes; IS92a,c,f, dots; IS92d,e, full lines; IS92b, short dashes. Radiative forcing sensitivities ($\partial Q/\partial C$) from Table 1, based largely on Shine et al. (1990); stratospheric ozone effects calculated using the method of Wigley and Raper (1992).

fers markedly between the 1990 IPCC "business as usual" scenario, SA90, and the later scenarios under the pre- and post-Copenhagen versions of the Montreal Protocol. In SA90, the no-ozone/ozone differential corresponds to continual increases in ozone depletion (which have a consequent negative radiative forcing effect). In IS92a,c,f (essentially a pre-Copenhagen scenario), the overall ozone effect is very small. The no-ozone and ozone curves for this scenario correspond to increased ozone depletion to around 2000, a slight improvement to 2030, increased depletion to 2060 (cf. the CFC11 and CFC12 emissions scenarios in Fig. 12.7), and ozone "repair" subsequently. In IS92d,e (a post-Copenhagen scenario), ozone depletion increases to a maximum around 2000, with subsequent recovery and continual ozone increases after that date. In terms of ozone depletion, the pre-Copenhagen situation represents a watershed, while the post-Copenhagen case (if global emissions accord with a scenario like IS92d,e) shows a substantial resolution of the problem. In terms of radiative forcing, however, the post-Copenhagen case shows that the stronger controls lead to greater radiative forcing due to the increasing ozone levels – although this is almost certainly the lesser of two evils.

The above ozone implications of the IPCC scenarios are consistent with current expectations based on much more detailed and complex chemical calculations. This consistency supports the realism of the simple calculation method used here in so far as the implied ozone changes are concerned. The overall radiative effect of these ozone changes is, however, still uncertain – perhaps by as much as ±50% (for the uncertainty in ΔQ_{OZ}). However, given that the total ΔQ under all IPCC92 scenarios is quite small (a maximum of 0.4W/m² by 2100), of which less than 50% is attributable to ozone effects (much less for IS92a,c,f), this uncertainty is of only minor importance.

In terms of climate change consequences, as noted earlier, the use of a global-mean figure in the above calculations is somewhat misleading because the radiative forcing changes are highly variable both spatially and seasonally. This is a scientifically interesting point, but it is of little practical significance simply because the overall ΔQ amounts are so small. Even if the lower-tropospheric climate-change signature of the spatially-variable halocarbon forcing effect were distinctively different from that due to the other greenhouse gases, it would be so small that it would be swamped by the uncertainties in the effects of the other gases and by the effects of natural climatic variability.

Sulphate Aerosol Effects

The possible importance of sulphate aerosols (arising from the oxidation of fossil-fuel-derived SO_2) has only recently been appreciated. It was because of this realization that Leggett et al. (1992) included SO_2 emissions in the IPCC92 scenarios. Aerosols have two different effects on the atmosphere's radiative balance. In clear sky conditions, they scatter a fraction of the incoming solar radiation back into space, promoting a surface cooling effect. They also affect the reflectivity of clouds by acting as cloud condensation nuclei (CCNs). In marine clouds, which are CCN deficient, more CCNs lead to more but smaller cloud droplets which, in turn, reflect more incoming solar radiation back to space. These two effects are referred to as the direct and indirect effects respectively. The former is thought to dominate, but the magnitudes of both effects are highly uncertain.

Here we follow Wigley and Raper (1992) in relating SO_2 emissions to radiative forcing. These authors use the three-dimensional model results of Charlson et al. (1991) for the direct effect, taking a Northern Hemisphere emission level of 64TgS/yr to correspond to a hemispheric-mean forcing of -1.07W/m². An estimate of about double this has been given by Charlson et al. (1992), based on a one-dimensional calculation, but Charlson (pers. comm., 1992) considers the earlier value to be superior. A recent re-calculation by Kiehl and Briegleb (1993), paralleling that of Charlson et al. (1991), gives a value of only around one-half the "best-guess" value used here, supporting our choice over the larger Charlson et al. (1992) estimate. For the indirect effect, we also follow Wigley and Raper (1992). This leads to an indirect ΔQ of around 20% of the direct value. As a guide to the overall uncertainty, we assume that the range is ±50% of the central estimate.

Emissions and best-guess radiative forcing projections under the IPCC92 scenarios are shown in Fig. 12.9. An important point to note with these results is that there is a marked hemispheric contrast in emissions, leading to a similarly marked ΔQ contrast. To date, more than 90% of fossil-fuel-derived SO_2 emissions have occurred in the Northern Hemisphere. Fig. 12.9 shows the radiative forcing differential in the year 2100 (SH minus NH) assuming an SH-NH split in future emissions of 10%–90%. The figures given are for the total differential, not just the change in the differential since 1990. (For 1990 global emissions of 75TgS/yr, as assumed here, the forcing differential is 1.20W/m², corresponding to forcings of –1.35W/m² and –0.15W/m² in the NH and SH respectively – as indicated on the zero ΔQ line in Fig. 129.)

Sulphate aerosol forcing is also highly spatially variable within each hemisphere, and varies with the time of year. The climate change signature of aerosol forcing is, therefore, likely to vary considerably with position and time of year. Thus, although it is true that the negative radiative forcing effect of aerosols offsets the positive radiative forcing effect of increasing greenhouse-gas concentrations in a global-mean sense, it would be entirely wrong to think that their climate consequences might cancel. Indeed, the spatially variable combined forcing effect is likely to be more disruptive to the climate system than the effect of greenhouse gases alone.

Relative Radiative Forcing Importance

The preceding sections give estimates of $\Delta Q(t)$ for the major anthropogenic greenhouse gases and for sulphate aerosols for each of the IPCC92 emissions scenarios. For CO_2 and N_2O, only best-guess results were given. Low, middle and high CO_2 results are given in the accom-

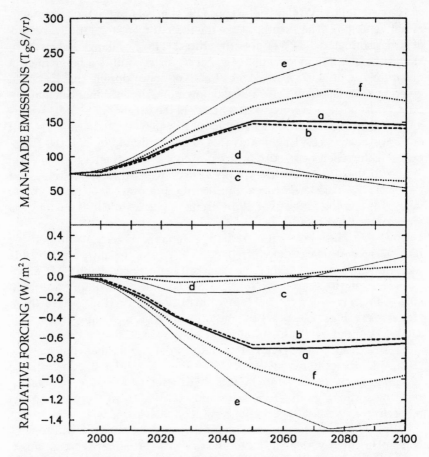

FIGURE 12.9: SO₂ emissions for the IPCC92 scenarios (IS92a–f) together with implied radiative forcing changes from 1990 (ΔQ). Forcing changes are best-guess values calculated using the method of Wigley and Raper (1992). They include both direct and indirect effects (dominated by the former).

panying chapter by Wigley (1994a). For N₂O, ΔQ uncertainties for any given emissions scenario are small; although it was noted that large uncertainties attend the emissions scenarios themselves. For CH₄, low, middle and high concentration results were given for the year 2100, but these were not interpreted in terms of ΔQ ranges. For halocarbons, results were given for two cases, with and without the effects of strato-spheric ozone changes. For sulphate aerosols, only best-guess results were shown, but it was noted that the ΔQ uncertainty for any given emissions scenario was at least ±50%. This section gives the best-guess total forcing for each IPCC emissions scenario, together with summaries

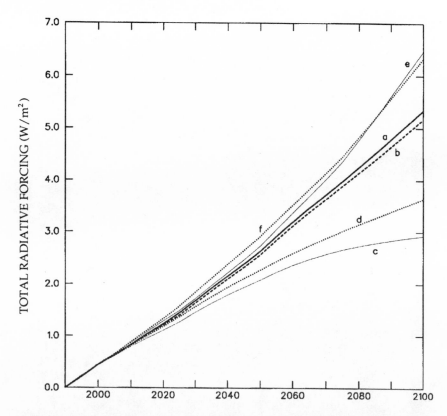

FIGURE 12.10: Total radiative forcing changes since 1990 for the IPCC92 emissions scenarios (updated from Wigley and Raper, 1992). Best-guess values are used for all concentration projections, ozone changes are accounted for in calculating the effects of halocarbons, stratospheric water vapour from methane oxidation is included, and best-guess aerosol forcing values are used.

of the gas-by-gas contributions in the years 2010, 2050 and 2100. Uncertainties are explored further in the next section.

Figure 12.10 shows the total radiative forcing changes for each scenario. These results are similar to those given by Wigley and Raper (1992, their Fig. 3). Minor differences (less than 0.1W/m² in the year 2100) occur because of improvements in the way halocarbon and aerosol forcings are calculated, a minor change in the model used to compute CH_4 concentration changes, and some minor changes in the carbon cycle model. Wigley and Raper noted the similarity between the forcing changes for the different scenarios out to around 2050, and their subsequent marked divergence. The former reflects lags between emissions and concentra-

TABLE 12.3: Radiative Forcing Changes (ΔQ) From 1990 (W/m^2) For the IPCC92 Emissions Scenarios (IS92a–f). All Values are Best-Guess Values. Halocarbon ΔQ Values Include the Effects of Stratospheric Ozone Changes, and Methane ΔQ Values Include the Contribution From Stratospheric Water Vapour Arising From Methane Oxidation. The Bottom Line Gives ΔQ Values to 1990 Based on Observed Concentration Changes. (Individual Values May Not Sum Precisely to Totals Due to Rounding Effects.)

Year	Scenario	CO_2	CH_4	N_2O	Halo-carbons	SO_2	Total	Tot./CO_2
2010	IS92a	.72	.18	.04	.07	-.16	.85	1.18
	IS92b	.70	.18	.04	.07	-.13	.86	1.23
	IS92c	.60	.14	.03	.07	-.03	.82	1.36
	IS92d	.62	.13	.03	.07	-.01	.85	1.37
	IS92e	.81	.20	.04	.07	-.21	.91	1.12
	IS92f	.77	.20	.04	.07	-.17	.91	1.18
2050	IS92a	2.44	.54	.15	.24	-.70	2.67	1.09
	IS92b	2.34	.54	.14	.24	-.66	2.60	1.11
	IS92c	1.59	.31	.12	.24	-.15	2.11	1.33
	IS92d	1.81	.26	.12	.24	-.03	2.41	1.33
	IS92e	3.06	.58	.16	.24	-1.18	2.87	0.94
	IS92f	2.81	.64	.16	.24	-.89	2.95	1.05
2100	IS92a	4.57	.84	.27	.35	-.65	5.38	1.18
	IS92b	4.36	.84	.26	.35	-.60	5.19	1.19
	IS92c	1.98	.25	.17	.34	.20	2.96	1.50
	IS92d	2.97	.21	.19	.38	.11	3.86	1.30
	IS92e	6.42	.97	.32	.38	-1.41	6.68	1.04
	IS92f	5.48	1.18	.31	.35	-.96	6.36	1.16
1990	Obs._Q	1.50	.56	.10	.02	-.75	1.42	0.94

tion changes, particularly for CO_2, while the latter is an expression of the increasing divergence between the emissions scenarios as one goes forward in time.

Table 12.3 shows individual (best-guess) ΔQ values from 1990 for each gas (total ΔQ for halocarbons) for the six scenarios in the years 2010, 2050 and 2100. It is clear from this table, as it is from the earlier sections, that CO_2 is by far the most important contributor. Next most important is SO_2, followed by CH_4 and then either N_2O or the halocarbons (depending on scenario). The relative importance of CO_2 depends critically on the scenario, increasing as one goes from IS92c to d to b to a to f to e (see last column of Table 12.3). This progression largely reflects changes in the negative forcing role of SO_2 emissions, which increasingly offset the CO_2 forcing changes in a global-mean sense as one progresses through the above scenario sequence (see Fig. 12.9). I note and stress again, however, the deceptive nature of these global-mean results, which hide large and important spatial differentials in the radiative forcing changes.

An alternative way to visualise the relative contributions of different gases to the overall radiative forcing changes is given in Fig. 12.11, which shows the percentage breakdown by decade for each scenario. The most striking results are:

1. The switch from a negative to a positive contribution from aerosols around the middle of next century as SO_2 emissions are assumed to peak and then decrease (cf. Fig. 12.9). This is particularly evident in IS92c where aerosols lead to a positive forcing contribution by 2100 that is larger than the CO_2 contribution (which, in this scenario, diminishes as concentration changes level off – see Fig. 12.2).
2. The change in methane from an enhancer to a reducer of ΔQ around 2060 in scenarios c and d; a result of the levelling off and subsequent reduction in concentrations in these scenarios at this time (see Fig. 12.3).
3. The substantial negative forcing contribution (largely in the NH) of aerosols in scenarios a, b, e and f out to 2050 (cf. Fig. 12.9, which gives the absolute ΔQ values).
4. The minor roles played by halocarbons and N_2O in all scenarios.

Radiative Forcing Uncertainties

In the previous section, only best-guess forcing changes were considered. Uncertainties in these changes arise for many reasons. Underlying all of these is the basic uncertainty in future emissions, characterized by the differences between the various scenarios. These different scenarios represent a range of results all based on the same background assumption of a maintenance of existing policies (see Leggett et al., 1992, for further details). The differences arise through different assumptions regarding: the implementation or otherwise of policies already proposed; projected population growth; projected growth in national economies; projected changes in per capita energy consumption and in emissions per unit of gross national product; projected resource availability; and so on.

For any given scenario, radiative forcing uncertainties arise for the following reasons: uncertainties in the relationships between emissions and concentration changes; possible emissions changes due to feedback effects associated with future climatic change; and uncertainties in radiative forcing changes per unit concentration change. An attempt has been made here to quantify the main factors in these categories. (For SO_2,

TABLE 12.4: Factors Leading to Uncertainties in Future Radiative Forcing Changes. The Numbers in the Right-Hand Column Show the Absolute Uncertainty Range in the Year 2100 (W/m^2).

FACTOR	RANGE (W/m^2)
1. Emissions scenario uncertainties (i.e., the differences between IS92a–f).	3.5
2. Carbon cycle modelling uncertainties. To around 2040, these exceed scenario uncertainties.[1]	1.0[2]
3. ΔE–ΔQ uncertainties for SO_2/aerosols (taken as ±50% here). Very dependent on scenario.	0.7[3]
4. ΔE–ΔQ uncertainties related to methane lifetime.	0.2
5. ΔE uncertainties due to methane emissions feedbacks.	0.2
6. ΔQ uncertainty for methane due to stratospheric water vapour.[4]	0.1[5]
7. Halocarbon uncertainties due to stratospheric ozone changes.	0.1[6]
8. Overall N_2O uncertainties, for any given scenario.	<0.1

[1] See accompanying paper by Wigley (1994a).

[2] Almost independent of scenario (see Wigley, 1994a).

[3] For IS92a, up to $1.4W/m^2$ for IS92e.

[4] Not discussed in text. Estimated at 50% of total water vapour ΔQ, which in turn is about 0.3/1.3 of the total methane ΔQ shown in Fig. 3.

[5] For IS92a; up to about 40% larger for IS92e.

[6] For IS92d,e; taken as 50% of the difference between the ΔQ results with and without ozone changes (see Fig. 8). Much smaller for IS92a,c and f.

ΔC–ΔQ uncertainty has been expressed via the ΔE ΔQ relationship. Because of the short lifetimes of SO_2 and sulphate aerosols, ΔC changes must parallel those in ΔE, so the ΔE–ΔQ link may be used as a proxy for the ΔC–ΔQ link.)

These results are summarized in Table 12.4, from which it is clear that the most important uncertainties are those associated with CO_2, SO_2 and CH_4. To give a better idea of the temporal evolution of these uncertainties, Fig. 12.12 shows, for each scenario, radiative forcing ranges arising from uncertainties in carbon (i.e., CO_2) and methane cycle modelling (i.e., associated with their ΔE–ΔC relationships) and uncertainties in aerosol forcing due to the ΔE–ΔQ link (taken as ±50%). For IS92a–d, CO_2 leads to the main uncertainty. For IS92e and f, however, the main uncertainty is that associated with SO_2.

Finally, uncertainties associated with scenario differences are considered gas-by-gas. These results are given in Table 12.5. As with all previous indicators, the greatest uncertainties here are those for CO_2, SO_2 and CH_4, in that order. Because of compensating effects with CO_2 and SO_2, the emissions projections of which are linked by their common

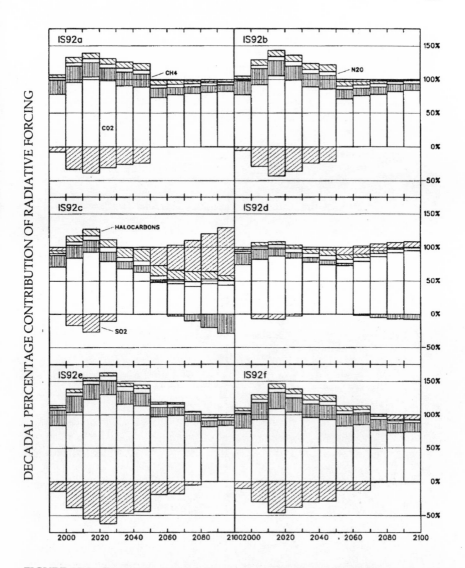

FIGURE 12.11: Percentage breakdown of decadal radiative forcing contributions for the IPCC92 emissions scenarios. Best-guess values are used for all concentration projections, ozone changes are accounted for in calculating the effects of halocarbons, stratospheric water vapour from methane oxidation is included, and best-guess aerosol forcing values are used.

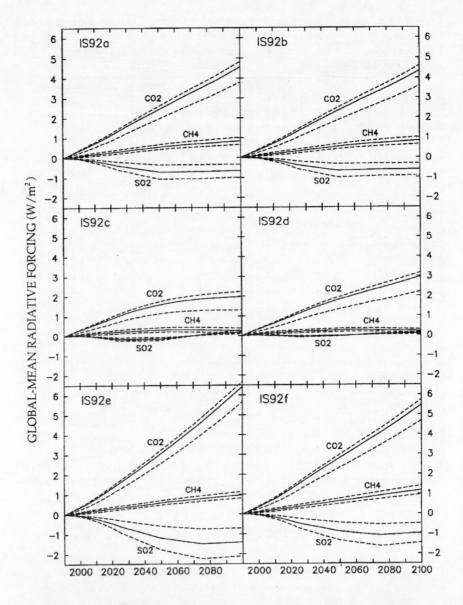

FIGURE 12.12: Radiative forcing uncertainties for CO_2, CH_4 and sulphate aerosols for the IPCC92 emissions scenarios. The uncertainties considered here are gas cycle modelling uncertainties for CO_2 and CH_4 and ΔE–ΔQ uncertainties for SO_2.

TABLE 12.5: Scenario Uncertainties In Radiative Forcing Changes For Individual Gases (W/m²). Shown are Lowest and Highest Values and Their Differences, the Inter-Scenario Ranges. (Extracted From Values Given in Table 12.3 of $\Delta Q(1990–2100)$).

Gas	Low/High ΔQ	Inter-scen. range
Carbon dioxide	1.98/6.42	4.44
Methane	0.21/1.18	0.97
Nitrous Oxide	0.17/0.32	0.15
Halocarbons	0.34/0.38	0.04*
Sulphate aerosols	-1.41/0.20	1.61
TOTAL	2.96/6.68	3.72

* This includes the effects of stratospheric ozone changes. For the direct halocarbon forcing (ignoring ozone) the inter-scenario range is 0.15W/m² (i.e., 0.21-0.36 W/m²).

fossil-fuel source, the scenario uncertainty associated with CO_2 emissions is actually greater than that for the total radiative forcing of all gases.

Conclusions

The main aim of this chapter has been to interpret the IPCC92 emissions scenarios for greenhouse gases and sulphur dioxide in terms of their implied concentration and radiative forcing changes over 1990–2100. From these changes, I have determined the relative importance of different gases as future forcing agents on the global climate system. In addition, I have attempted to assess and quantify the major uncertainties in these factors. This is the first time that anyone has attempted to give a comprehensive, quantitative assessment of these uncertainties. The ranges of uncertainty given are not meant to span the full spectrum of possibilities – indeed, the assessment of uncertainty is subject to its own substantial uncertainties. Nevertheless, I consider it relatively unlikely, for any given scenario, that the results would lie outside the estimated uncertainty ranges. The uncertainty ranges given correspond roughly to 90% confidence bands.

Knowing the uncertainties is a pre-requisite for defining the task of reducing them. In some cases, uncertainty reduction poses formidable difficulties, while in others substantial gains could be made through relatively modest research efforts. The research priorities should be determined by both the magnitude of the uncertainty and its likely reduction per unit of effort. Expressing all uncertainties in common units (viz. radiative forcing) is an essential starting point in this prioritization.

From Table 12.4, it is clear that the largest uncertainties arise from the emissions scenarios themselves if one considers the situation in the second half of the 21st century. These uncertainties arise largely from the effects of CO_2 and SO_2, which, in turn, reflect uncertainties in the future use of fossil fuels as energy sources. Out to around 2040, however, as noted by Wigley and Raper (1992), ΔQ uncertainties arising from uncertainties in future emissions are relatively small (see Fig. 12.10). Prior to this date, uncertainties in the carbon cycle and, in some scenarios, in the relationship between SO_2 emissions and aerosol radiative forcing are the most important. In both of these areas, accelerated (but appropriately targetted) research over the next decade, which would have a relatively small cost, would lead to considerable improvements.

The next level of priority, somewhat below the above, should be research into methane chemistry and climate-related methane emissions feedbacks. On the long time scale (i.e., beyond 2100), the latter could become considerably more important than in the assessment made here. Nevertheless, there appears to be little urgency for research in this area, especially given that the only way to control these feedbacks is by limiting the extent of global warming, the uncertainties in which are determined largely by other factors.

At a lower priority level, but qualitatively similar in that the long-term implications may be relatively greater than those over the 100-year horizon considered here, is the issue of N_2O emissions; specifically, current budget uncertainties and possible climate-related feedbacks. Similar research priorities can be placed on improving our knowledge of the enhancement of methane ΔQ through its oxidation product, stratospheric water vapour; and on improving our understanding of how stratospheric ozone changes alter the overall radiative forcing effects of chlorine- and bromine-containing halocarbons. Progress in both of these areas is likely to be made over the next few years.

Numerous scientific issues have been neglected in this review, some of which may have radiative forcing consequences similar to the items considered in the previous two paragraphs. Most notable of these is tropospheric ozone. In the upper troposphere, this is a strong greenhouse gas, and improved understanding is needed both of past concentration changes and possible future changes. Another indirect consequence of methane emissions, their effect on lower stratospheric ozone (more methane means more ozone), also requires better quantification.

Finally, I need to stress two points. First, the research priorities assigned above are based largely on the consequences improved knowledge might have in terms of reducing future ΔQ uncertainties.

Relatively little weight has been placed directly on the intrinsic scientific value of the research. This is not because I believe this to be a secondary consideration, but because I chose at the outset a goal-oriented approach (i.e., reducing ΔQ uncertainties) for the primary terms of reference. Second, I have focussed attention on global-mean radiative forcing because of its conceptual simplicity and unifying value. For some of the gases considered here, the radiative forcing changes are highly spatially non-uniform. Particularly for SO_2, elucidating the climate response to such non-uniform forcing changes should be a topic of considerable priority.

References

Charlson, R. J., Langner, J., Rodhe, H., Leovy, C. B. and Warren, S. G. 1991. "Perturbation of the Northern Hemisphere radiative balance by backscattering from anthropogenic sulfate aerosols." *Tellus* 43AB, 152 163.

Charlson, R. J., Schwartz, S. E., Hales, J.M., Cess, R. D., Coakley, J. A., Hansen, J. E. and Hofmann, D. J. 1992. "Climate forcing by anthropogenic aerosols." *Science* 255, 423 430.

Hansen, J., Lacis, A., Rind, D., Russell, L., Stone, P., Fung, I., Ruedy, R. and Lerner, J. 1984. "Climate sensitivity analysis of feedback mechanisms," in, *Climate Processes and Climate Sensitivity* (J. Hansen and T. Takahashi, Eds.). *Geophysical Monograph 29*, American Geophysical Union, Washington, D.C., pp.130 163.

Isaksen, I. S. A., Ramaswamy, V., Rodhe, H. and Wigley, T. M. L. 1992a. In, *Climate Change 1992: The Supplementary Report to the IPCC Scientific Assessment* (J.T. Houghton, B.A. Callander and S.K. Varney, Eds.). Cambridge: Cambridge University Press, pp.47-67.

Isaksen, I. S. A., Brühl, C., Molina, M., Schiff, H., Shine, K. and Stordal, F. 1992b. *An Assessment of Role of CF$_4$ and C$_2$F$_6$ as Greenhouse Gases.* CICERO Policy Note 1992:6, Oslo, 30pp.

Kiehl, J. T. and Briegleb, B. P. 1993. "The relative role of sulfate aerosols and greenhouse gases in climate forcing." *Science* 260, 311–314..

Leggett, J., Pepper, W. J. and Swart, R. J. 1992. "Emissions scenarios for IPCC: An update," in, *Climate Change 1992: The Supplementary Report to the IPCC Scientific Assessment* (J.T. Houghton, B.A. Callander and S.K. Varney, Eds.). Cambridge: Cambridge University Press, pp.69-96.

Leuenberger, M. and Siegenthaler, U. 1992. "Ice-age atmospheric concentration of nitrous oxide from an Antarctic ice core." *Nature* 360, 449-451.

Manabe, S. and Wetherald, R. T. 1980. "On the distribution of climate change resulting from an increase in CO_2 content of the atmosphere." *Journal of the Atmospheric Sciences* 37, 99-118.

Osborn, T. J. and Wigley, T. M. L. 1992. "A simple model for estimating methane concentration and lifetime variations." *Climate Dynamics* (in press)

Ramaswamy, V., Schwarzkopf, M. D. and Shine, K. P. 1992. "Radiative forcing of climate from halocarbon-induced global stratospheric ozone loss." *Nature* 355, 810-812.

Rogers, J. E. and Whitman, W. B., Eds. 1991. *Microbial Production and Consumption of Greenhouse Gases: Methane, Nitrogen Oxides and Halomethanes*, American Society for Microbiology, Washington, D.C., 298 pp.

Shine, K. P., Derwent, R. G., Wuebbles, D. J. and Morcrette, J. J. 1990. "Radiative forcing of climate," in, *Climate Change: The IPCC Scientific Assessment* (J.T. Houghton, G. J. Jenkins and J. J. Ephraums, Eds.). Cambridge: Cambridge University Press, pp.41-68.

Wang, W. C., Dudek, M. P., Liang, Z. Z. and Kiehl, J. T. 1991. "Inadequacy of effective CO_2 as a proxy in simulating the greenhouse effect of other radiatively active gases." *Nature* 350, 573-577.

Watson, R. T., Rodhe, H., Oeschger, H. and Siegenthaler, U. 1990. "Greenhouse gases and aerosols," in, *Climate Change: The IPCC Scientific Assessment* (J.T. Houghton, G.J. Jenkins and J.J. Ephraums, Eds.). Cambridge: Cambridge University Press, pp.1 40.

Watson, R. T., Meira Filho, L. G., Sanhueza, E. and Janetos, A. 1992. "Sources and sinks," in, *Climate Change 1992: The Supplementary Report to the IPCC Scientific Assessment* (J.T. Houghton, B.A. Callander and S.K. Varney, Eds.). Cambridge: Cambridge University Press, pp.25-46.

Wigley, T. M. L. 1994a. How important are carbon cycle model uncertainties? (this volume).

Wigley, T. M. L. 1994b. "Balancing the carbon budget: implications for projections of future carbon dioxide concentration changes." *Tellus* (in press).

Wigley, T. M. L. and Raper, S. C. B. 1992. "Implications for climate and sea level of revised IPCC emissions scenarios." *Nature* 357, 293-300.

Support from the U.S. Dept. of Energy (Grant No. DE-FG02-86ER60397) is gratefully acknowledged. This work was carried out while the author was at the Climatic Research Unit, University of East Anglia, Norwich, U.K.

About the Book

Drawn from issues discussed at the 1992 Earth Summit, this volume of essays addresses the most strategic questions and challenges to scientists and policymakers on the important subject of climate change. Sponsored by CICERO (Center for International Climate and Energy Research–Oslo), a policy research foundation of the University of Oslo, the book features an international cast of environmental, science, and policy experts who assess implications, strengths, and limitations of the Rio Convention while considering how best to meet the challenge that atmospheric pollution poses worldwide. Issues covered include: the spread of beneficial technology and the competence required in developing countries; improving inventories of greenhouse gases; calculating the most effective mix of measures, nationally and regionally; meeting future energy needs for countries with different economic structures while limiting emissions of greenhouse gases.